Annals of Mathematics Studies

Number 48

LECTURES ON MODULAR FORMS

BY

R. C. Gunning

NOTES BY

ARMAND BRUMER

PRINCETON, NEW JERSEY
PRINCETON UNIVERSITY PRESS
1962

L. C. Card 62-7403
ISBN 0-691-07995-1

AMS 1968: 10D05, 30A58

Printed in the United States of America

INTRODUCTION

These are notes based on a course of lectures given at Princeton University during the Spring Semester of 1959, on the subject of modular forms of one complex variable. There has been a resurgence of interest in this subject recently, caused no doubt by the new results and techniques of Selberg and of Eichler; but there does not exist any introductory text which covers the background for these current developments, particularly the relevant parts of the works of Hecke and Petersson. The lectures were designed to fill in this gap to some extent. I have not attempted to give a comprehensive discussion of these topics, but rather to give a brief survey illustrating the techniques and problems of some aspects of the subject. For this reason only the simpler cases were treated – modular forms of even weights without multipliers, the principal congruence subgroups, the Hecke operators for the full modular group alone.

Princeton, New Jersey
November, 1961

R. C. Gunning

CONTENTS

CHAPTER I. GEOMETRICAL BACKGROUND....................................

§1. Linear Fractional Transformations....................

§2. The Modular Group..............................

§3. The Principal Congruence Subgroups..................

§4. The Riemann Surfaces Associated to Subgroups of the Modular Group.................................... 1

§5. Examples.. 1

CHAPTER II. MODULAR FORMS.. 1

§6. Introduction...................................... 1

§7. A Review of Some Function Theory on Riemann Surfaces. 2

§8. The Dimension of the Space of Modular Forms.......... 2

CHAPTER III. POINCARÉ SERIES...................................... 2

§9. Construction of Modular Forms........................ 2

§10. The Petersson Inner Product......................... 3

§11. Completeness of Poincaré Series..................... 3

§12. The Fourier Coefficients of Cusp Forms.............. 4

CHAPTER IV. EISENSTEIN SERIES.................................... 4

§13. Construction of the Eisenstein Series............... 4

§14. The Fourier Coefficients of Eisenstein Series........ 4

§15. The Modular Forms for the Modular Group............. 5

CHAPTER V. MODULAR CORRESPONDENCES............................ 5

§16. The Hecke Operators................................. 5

§17. Hecke Operators and Fourier Coefficients............ 6

§18. Arithmetic Properties of the Fourier Coefficients.... 6

CHAPTER VI. QUADRATIC FORMS...................................... 7

§19. Introduction.. 7

§20. The Generalized Jacobi Inversion Formula............. 7

§21. Representations by Sums of Squares.................. 7

§22. Even Integral Quadratic Forms........................ 7

§23. Arithmetic Applications............................. 8

LECTURES ON MODULAR FORMS

CHAPTER I:

GEOMETRICAL BACKGROUND

§1. Linear Fractional Transformations

In this section we review some results about linear fractional transformations which will be needed. For proofs and additional details the reader is referred to [1].*

The only conformal automorphisms of the Riemann sphere onto itself are the linear fractional transformations:

$$T:\ z \longrightarrow \frac{az+b}{cz+d},$$

where $\begin{pmatrix} a & b \\ c & d \end{pmatrix}$ is a matrix of complex coefficients having determinant one. The set of all linear fractional transformations form a group which is isomorphic to the group of complex 2×2 matrices of determinant 1 (denoted by $SL(2,Z)$) divided by its center, $\pm \begin{pmatrix} 1 & 0 \\ 0 & 1 \end{pmatrix}$.

Each linear fractional transformation except the identity transformation has at most two fixed points, and at least one fixed point. The group of all linear fractional transformations acts transitively on the sphere; indeed there is a unique linear fractional transformation which sends any given ordered triple of distinct points, into any other given such triple. A linear fractional transformation also transforms circles on the Riemann sphere into circles.

We shall consider henceforth only linear fractional transformations with real coefficients; these map the upper half plane onto itself. Any such transformation is one of the following three types:

(1) Elliptic transformation. This transformation has two fixed

---

* Square brackets refer to the bibliography at the end of the Chapter.

points $\zeta, \bar{\zeta}$, with ζ in upper half plane ($\bar{\zeta}$ indicates the complex conjugate of ζ). After a change of variable sending $\{\zeta, \bar{\zeta}\}$ into $\{0, \infty\}$, the transformation takes on the normal form:

$$w' = Kw \ , \ K = e^{i\theta} \ ;$$

(i.e., it is a rotation about 0 through an angle of θ).

(2) Hyperbolic transformation. This transformation has two fixed points on the real axis. When these are sent into 0 and ∞ by a suitable change of variable the transformation takes on the normal form:

$$w' = Kw \ , \ K > 0 \ ;$$

(i.e., it is a dilation of magnitude K with center at the origin).

(3) Parabolic transformation. This transformation has only one fixed point, either at ∞ or on the real line. By sending the fixed point to ∞, if necessary, the transformation takes on the normal form:

$$w' = w + c \ .$$

It is possible to determine quite readily the type of any transformation

$$T: z \longrightarrow \frac{az + b}{cz + d} \ ,$$

1. If $c = 0$, then T is parabolic.
2. If $c \neq 0$, then:

$$|a + d| > 2 \Longleftrightarrow T \text{ hyperbolic}$$
$$|a + d| = 2 \Longleftrightarrow T \text{ parabolic}$$
$$|a + d| < 2 \Longleftrightarrow T \text{ elliptic.}$$

§ 2. The Modular Group

We shall be interested in analytic functions on the upper half plane which have simple invariance properties under some groups of linear fractional transformations; the groups of particular interest are those which have simple arithmetic definitions.

Definition: The modular group, Γ, is the group of linear fractional transformation:

$$L: z \longrightarrow \frac{az + b}{cz + d} \ , \ ad - bc = 1,$$

where a,b,c,d are integers. (There are actually two groups which can be considered here, and a little care must be taken to keep them separate. First is the group $\Gamma' = SL(2,Z)$ of 2×2 matrices with integer coefficients and determinant 1; this is the homogeneous modular group. The matrices $\begin{pmatrix} a & b \\ c & d \end{pmatrix}$ and $\begin{pmatrix} -a & -b \\ -c & -d \end{pmatrix}$ clearly determine the same linear fractional transformation however. Therefore the group of distinct linear fractional transformations is the quotient group $\Gamma \cong \Gamma'/(\pm I)$ where $I = \begin{pmatrix} 1 & 0 \\ 0 & 1 \end{pmatrix}$; this is the inhomogeneous modular group.)

Using the criterion given at the end of §1, we see that the elliptic transformations of Γ occur when $a + d = 0$ or when $a + d = \pm 1$. A simple calculation shows that the first case leads to a transformation L for which $L^2 = I$ while in the second case $L^3 = I$.

Definition: Let H be the upper half-plane. Two points z_1, z_2 in H are equivalent under a group G of transformations of the upper half plane (written $z_1 \sim_G z_2$) if there is a transformation T in G such that $z_1 = Tz_2$. This is clearly an equivalence relation. A fundamental domain for the group G is an open set D which does not contain any pairs of distinct equivalent points and whose point set closure contains at least one point from each equivalence class.

It follows from the definition that the transforms of the fundamental domain D by elements of the group cover the entire upper half plane, and that two transforms of D whose intersection contains an open set must coincide.

Our first task is to find a fundamental domain for the modular group Γ.

Lemma 1: For a fixed point z in H there are only a finite number of pairs of integers (c,d) such that

$$|cz + d| \leq 1 .$$

Proof: Let (c,d) be such a pair; then

$$|cz + d|^2 = (cx + d)^2 + c^2y^2 ,$$

so that

$$c^2y^2 \leq (cx + d)^2 + c^2y^2 \leq 1.$$

Since z is in H, $y > 0$; then

$$|c| \leq \frac{1}{y}$$

and there are hence only a finite number of possible values for c. For any such value of c the equation

$$(cx + d)^2 + c^2y^2 \leq 1$$

shows that there are only a finite number of possible values of d.

It is convenient to call $y = \text{Im } z$ the <u>height</u> of $z = x + iy$.

<u>Lemma 2</u>: Among the transforms $\{Tz\}$ of a point z in H there are only a finite number with heights larger than the height of z.

<u>Proof</u>: For any z in H and T in Γ,

$$Tz = \frac{az + b}{cz + d} = \frac{az + b}{cz + d}\,\frac{c\bar{z} + d}{c\bar{z} + d} = \frac{\text{Real} + i(ad - bc)\text{Im } z}{|cz + d|^2}$$

so that

$$\text{Im } Tz = \frac{\text{Im } z}{|cz + d|^2} \, .$$

The desired result then follows from Lemma 1.

Lemma 2 suggests that we select from each equivalence class an element of maximum height, i.e., a point z such that $|cz + d| \geq 1$, for all integer pairs c,d. Since the translation $T: z \longrightarrow z + 1$ is in Γ we can further assume that the fundamental domain lies in the strip $|\text{Re } z| = |x| \leq \frac{1}{2}$. The second normalization does not destroy the first.

<u>Theorem 1</u>: A fundamental domain for Γ is the set

$$D = \{z \text{ in } H| \ |\text{Re } z| < \tfrac{1}{2} \text{ and } |z| > 1\}.$$

<u>Proof</u>: We first show that D is the same as the set

$$D_1 = \{z \text{ in } H| \ |\text{Re } z| < \tfrac{1}{2} \text{ and } |cz + d| > 1\}$$

for all pairs of integers c,d except $c = d = 0$. Setting $c = 1$, $d = 0$, shows that $D_1 \subseteq D$. Conversely suppose that z is in D; then
$|cz + d|^2 = (cx + d)^2 + c^2y^2 = c^2(x^2 + y^2) + 2cdx + d^2 > c^2 - cd + d^2 \geq 1$
for all pairs of integers c,d except the pair $c = d = 0$. Now from our earlier remarks it follows that the closure of D contains at least one point from each equivalence class under the modular group Γ. We show that the only pairs of points of the closure of D which are equivalent under Γ

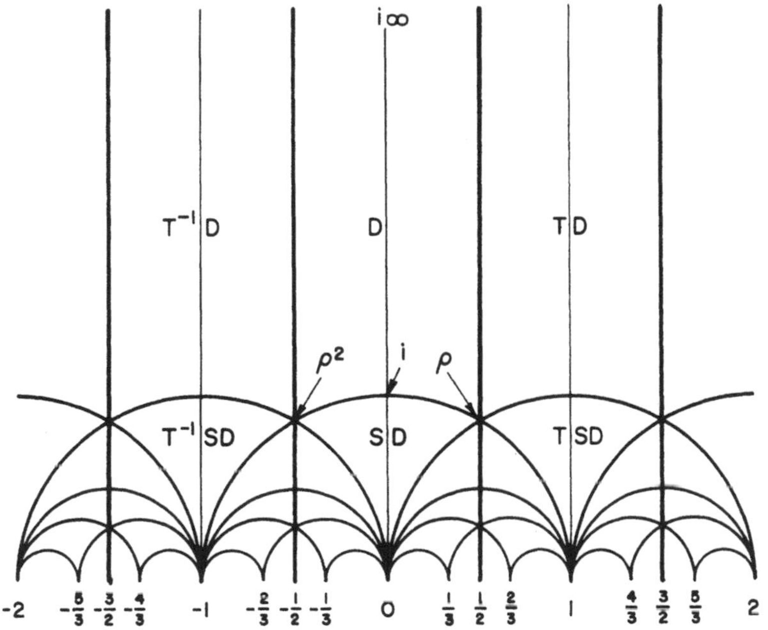

Figure 1.

are the pairs of points of the boundary of D which coincide upon reflexion about the line $x = 0$ (see Figure 1); these points are identified by the transformations $T: z \longrightarrow z + 1$ and $S: z \longrightarrow -\frac{1}{z}$. Suppose z, z' are in D and $z \sim_\Gamma z'$, say $z' = Lz$; then $\operatorname{Im} z = \operatorname{Im} Lz$, so that

$$1 = |cz + d| = (cx + d)^2 + c^2y^2 \geq c^2 + d^2 - cd \geq 1,$$

hence either $c = 0$, $d = \pm 1$ or $d = 0$, $c = \pm 1$. A simple calculation shows that the first case leads to $L = T$, and the second to $L = S$; these are the identifications already mentioned, thus concluding the proof.

The only fixed points of transformations of Γ which lie in $\overline{D}$ are the points i, ρ, and ρ^2, where $\rho = e^{\pi i/3}$ and $i = \sqrt{-1}$. These points are fixed points under the elliptic transformations $S: z \longrightarrow -\frac{1}{z}$ (of order 2), $TS: z \longrightarrow \frac{z-1}{z}$ (of order 3) and $ST: z \longrightarrow \frac{-1}{z+1}$ (of order 3), respectively.

Remark: The two transformations $T: z \longrightarrow z + 1$ and $S: z \longrightarrow -\frac{1}{z}$

generate the inhomogeneous modular group. They satisfy the relations $S^2 = (TS)^3 = I$, and all other relations are consequences of these. It is indeed easy to see that $S^2 = (TS)^3 = I$, but it is not so obvious that these are, essentially, the only relations. Since this result is not used directly here, no proof will be given; but see [2]. To show that S and T generate the modular group, let G be the subgroup of Γ generated by T and S. Note first that the transforms of D by elements of G cover H, the upper half plane. For let z be any point of H; after applying T or T^{-1} sufficiently often, the transform $z_1 = T^m z$ will satisfy $|\mathrm{Re}\, z_1| \leq \frac{1}{2}$. If z_1 is in D we are through; otherwise $|z_1| < 1$ and the height of Sz_1 is <u>strictly</u> <u>larger</u> than the height of z_1 (which is equal of course to the height of z). Repeating the construction now with the point Sz_1, and continuing this procedure, some transform of z by an element of G must eventually lie in D, since by Lemma 2, there are only a finite number of heights of transforms of z which are strictly larger than the height of z. Now let L be a transformation in Γ and z be some interior point of D. As above, there is a transformation g in G such that $gz = Lz$. Then $L^{-1}g$ is in Γ and leaves z fixed; but since z is left fixed only by the identity it follows that $g = L$, hence that $G = \Gamma$, as desired.

Consider H/Γ, the set of equivalence classes of points of the upper half-plane H under the modular group Γ. We topologize H/Γ with the strongest topology under which the natural map $\tau\colon H \longrightarrow H/\Gamma$ is continuous; H/Γ is simply $\bar{D}$ with proper identifications along the boundary. We should like to put an analytic structure on H/Γ. Around any point z in D which is not a fixed point we can draw a small disc not containing any fixed points which is mapped homeomorphically onto an open neighborhood of τz in H/Γ. This defines a parametric disc about τz. The elliptic fixed points must be treated separately. There are only two equivalence classes of such points, i and $\rho = e^{\pi i/3}$. The transformation S is a rotation of period 2 with fixed point i. We use a "half disc" N about i as a parametric disc (see Figure 2); the two "radii" are identified by S. To map N homeomorphically onto a disc, we first transform S into normal form by sending i into 0 and -i into ∞, by means of a linear

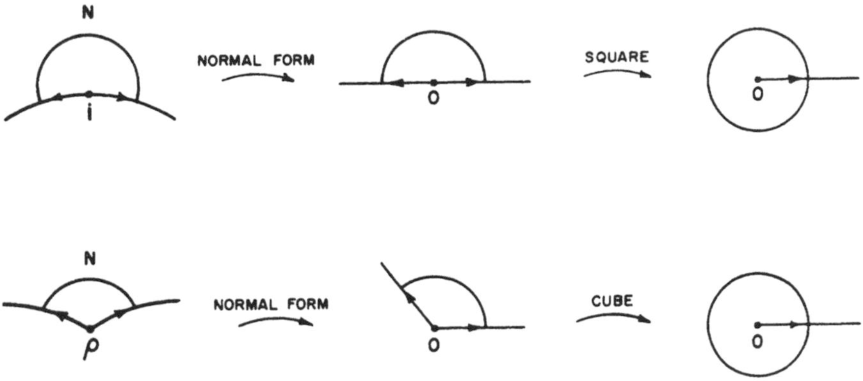

Figure 2.

fractional transformation. The transform of N is then a true half-disc, where the two bounding radii are to be identified. By sending every element into its square we get a full disc, which can be used as a parametric disc about i. About ρ (see Figure 2) we use the same procedure; since TS is a rotation through $120°$, however, we must send every element into its cube, rather than its square, in the final step.

We compactify H/Γ by adding the point at $i\,\infty$; to put a complex structure on the compactification, we must find a parametric disc about $i\,\infty$. The set $\{z \text{ in } H \mid \text{Im } z > 1\}$ is mapped by $t: z \longrightarrow e^{2\pi i z}$ onto the punctured disc $|t| < e^{-2\pi}$. Also, for fixed x, as y tends to $+\infty$, $\arg(t(z))$ remains constant while $|t(z)|$ approaches 0. Finally, two points z, z' of the set are mapped into the same point only if $z' = z + m$ for some integer m; but these points are the same in H/Γ. Therefore we can compactify H/Γ by adding the point $t = 0$ in this parametric disc.

The compact Riemann surface $\overline{H/\Gamma}$ is seen to be a sphere, either by noticing that it is simply connected and applying the uniformization theorem, or else by using the natural triangulation (see Figure 1). We have thus proved:

<u>Theorem 3</u>: The identification space H/Γ, when compactified by adding the point $i\,\infty$, can be given a natural analytic structure under which it is a compact Riemann surface of genus 0.

§3. The Principal Congruence Subgroups.

In this section we shall study an important class of subgroups of the modular group. These groups will arise naturally in our applications to quadratic forms.

Definition: Let Z be the group of the integers, and Z_q be the integers modulo q. The natural homomorphism $\varphi: Z \longrightarrow Z_q$ induces a homomorphism $\tilde{\varphi}: SL(2,Z) \longrightarrow SL(2,Z_q)$ defined by

$$\tilde{\varphi}\begin{pmatrix} a & b \\ c & d \end{pmatrix} = \begin{pmatrix} \varphi(a) & \varphi(b) \\ \varphi(c) & \varphi(d) \end{pmatrix} .$$

The kernel of this map, Γ'_q, is called the (homogeneous) principal congruence subgroup of level q. (The word "Stufe" is frequently used in place of level.)

Theorem 2: Γ'_q is a normal subgroup of Γ', and $\Gamma'/\Gamma'_q \approx SL(2,Z_q)$.

Proof: It is sufficient to show that the map $\tilde{\varphi}$ is onto. Let $\begin{pmatrix} a & b \\ c & d \end{pmatrix}$, $ad - bc \equiv 1 \pmod q$, represent a matrix in $SL(2,Z_q)$. We can write the determinant condition in the form $ad - bc - mq = 1$ for some integer m; hence $(c,d,q) = 1$.* We can therefore find an integer n such that $(c,d + nq) = 1$, and can thus assume that $(c,d) = 1$. Consider the matrix

$$\begin{pmatrix} a + eq & b + fq \\ c & d \end{pmatrix} .$$

Its determinant is $ad - bc + q(ed - fc) = 1 + (m + ed - fc)q$. Since $(c,d) = 1$, there exist integers e,f such that $m = fc - ed$. This then provides a matrix in $SL(2,Z)$ representing the given matrix in $SL(2,Z_q)$.

Our next task will be to compute the index $\nu(q)$ of Γ'_q in Γ', or, equivalently, the order of $SL(2,Z_q)$. For this purpose we introduce the following concept:

Definition: A pair of integers c,d is called primitive mod q, if $(c,d,q) = 1$. The number of incongruent primitive pairs mod q mod q will be denoted by $\lambda(q)$.

Lemma 3: The second rows of matrices representing elements of $SL(2,Z_q)$ are precisely the primitive pairs mod q.

Proof: If $\begin{pmatrix} a & b \\ c & d \end{pmatrix}$ represents an element of $SL(2,Z_q)$, then

---

* We are using the usual notation for the greatest common divisor.

d - bc ≡ 1 (mod q), or ad - bc - mq = 1; so that c,d is a primitive air mod q. Conversely, if (c,d,q) = 1 there exist integers a,b,m such hat ad - bc - mq = 1; thus $\begin{pmatrix} a & b \\ c & d \end{pmatrix}$ represents an element of $SL(2,Z_q)$.

Lemma 4: For a fixed primitive pair c,d of integers mod q there re q incongruent pairs of integers a,b mod q such that ad - bc ≡ 1 mod q); or, equivalently, there are q distinct elements of the form $\begin{pmatrix} a & b \\ c & d \end{pmatrix}$ in $SL(2,Z_q)$.

Proof: The verification is straightforward and will be left to the eader.

Lemma 5: $\lambda(q)$ is a multiplicative function of q; i.e., if $q_1,q_2) = 1$, then $\lambda(q_1)\,\lambda(q_2) = \lambda(q_1q_2)$.

Proof: Let $\{\gamma_i,\delta_i\}$ be a primitive pair for q_i (i = 1,2). Then $\gamma_1q_2 + \gamma_2q_1,\ \delta_1q_2 + \delta_2q_1\}$ is a primitive pair for q_1q_2 since $(q_1,q_2) = 1$. lso, incongruent pairs for q_1 and q_2 lead to incongruent pairs for q_1q_2. hus $\lambda(q_1)\,\lambda(q_2) \leq \lambda(q_1q_2)$. Conversely let $\{\gamma,\delta\}$ be a primitive pair od q_1q_2; then $\{\gamma,\delta\}$ is a primitive pair mod q_1 and mod q_2. Also, ince $(q_1,q_2) = 1$, incongruent pairs mod q_1q_2 cannot give rise to conruent ones both mod q_1 and mod q_2. Thus $\lambda(q_1q_2) \leq \lambda(q_1)\,\lambda(q_2)$.

Lemma 6: If p is a prime, $\lambda(p^k) = p^{2k}(1 - \frac{1}{p^2})$.

Proof: There are $p^k(1 - \frac{1}{p})$ incongruent integers c mod p^k such hat $(c,p) = 1$. For any one of these, any of the p^k incongruent values f d will give a primitive pair. Since these pairs are all incongruent od p^k there are $p^{2k}(1 - \frac{1}{p})$ such pairs. Now there are p^{k-1} values of such that $(c,p) = p$. To these correspond $p^k(1 - \frac{1}{p})$ values of d which re incongruent and such that $(d,p) = 1$. These lead to $p^{2k-1}(1 - \frac{1}{p})$ rimitive pairs. Adding the two cases gives $\lambda(p^k) = p^{2k}(1 - \frac{1}{p^2})$.

Theorem 3: The index $\nu'(q)$ of Γ'_q in Γ' is

$$\nu'(q) = q^3 \prod_{p \mid q} (1 - \frac{1}{p^2}).$$

Proof: This follows from the last four lemmas.

Let $\Gamma^*_q = \left\{ \begin{pmatrix} a & b \\ c & d \end{pmatrix} \text{ in } \Gamma' \;\middle|\; \begin{pmatrix} a & b \\ c & d \end{pmatrix} \equiv \pm I \pmod q \right\}$.

hen Γ^*_q is a normal subgroup of Γ. Let $\Gamma_q \cong \Gamma^*_q/(\pm I)$ be the corresponding normal subgroup of the inhomogeneous modular group Γ; and let $\nu(q) = [\Gamma : \Gamma_q]$ be the index of Γ_q in Γ. If $q = 2$ then $\Gamma_q^* = \Gamma_2'$ since $I \equiv -I (\text{mod } 2)$. Thus $\nu(2) = \nu'(2) = 8(1 - \frac{1}{4}) = 6$. If $q > 2$ then $[\Gamma_q^* : \Gamma_q] = 2$, so that $\nu(q) = \frac{1}{2}\nu'(q) = \frac{1}{2} q^3 \prod_{p|q} (1 - \frac{1}{p^2})$. Finally, if $q = 1$ then $\nu(1) = 1$.

§4. The Riemann Surfaces Associated to Subgroups of the Modular Group.

Let $G \subseteq \Gamma$ be a subgroup of the modular group of finite index μ. We shall find a fundamental domain for G, which can be compactified and made into a Riemann surface as in §2; and we shall then compute the genus of that Riemann surface, which will be used in later applications of the Riemann-Roch theorem.

Theorem 4: Let G be a subgroup of index μ in Γ and select coset representatives $T_1, \ldots, T_\mu$ so that $\Gamma = GT_1 \cup GT_2 \cup \cdots \cup GT_\mu$. If D is a fundamental domain for Γ then

$$D_G = T_1 D \cup T_2 D \cup \cdots \cup T_\mu D$$

is a fundamental domain for G.

Proof: Clearly the transforms of D_G by elements of G cover the upper half plane. If $SD_G \cap D_G$ contained an open set, that set would in turn contain a transform of D; but then $ST_i D = T_j D$, which would imply $ST_i = T_j$, a contradiction.

Now the quotient space H/G can be given an analytic structure, just as H/Γ was in §2. As for the compactification, there may be real parabolic vertices as well as the infinite parabolic vertices, but all are treated in basically the same way.

As mentioned earlier there is a natural triangulation of H/Γ (see Figure 1) in which the fixed points are the vertices and every 1-simplex connects two fixed points. This triangulation of H/Γ induces a triangulation on $\overline{H/G}$. Also $\overline{H/G}$ is a compact Riemann surface. We compute its genus by means of the Euler characteristic formula

$$\chi = 2 - 2p = \sigma_0 - \sigma_1 + \sigma_2 , \tag{1}$$

where χ is the Euler characteristic, p is the genus, and σ_k is the number of k-simplexes in the triangulation (see [3]). In the natural triangulation of $\overline{H/G}$, σ_0 is the number of images of elliptic and parabolic points of Γ. It is convenient to write $\sigma_0 = \lambda_i + \lambda_\rho + \lambda_\infty$, where λ_i $(\lambda_\rho, \lambda_\infty)$ is number of vertices equivalent to i (ρ, ∞). Let $\rho_1, \ldots, \rho_{\sigma_0}$ be the vertices of the triangulation, the first λ_i being equivalent to i, and the next λ_ρ being equivalent to ρ. To find out how many 1-simplexes meet at a typical vertex p_k we distinguish various cases:

(a) If p_k is equivalent to i then two or four 1-simplexes meet at p_k according as it is a fixed point for G or not.

(b) If p_k is equivalent to ρ then two or six 1-simplexes meet at p_k according as it is a fixed point for G or not.

(c) If p_k is equivalent to ∞ then, if it compactifies n fundamental domains (for Γ), $2n$ 1-simplexes meet there.

In summary, an even number of 1-simplexes, say $2n_k$, meet at p_k.

<u>Theorem 5</u>: In the notation above the genus of $\overline{H/G}$ is given by:

$$p = 1 + \frac{1}{2}(\mu - \sigma_0) \;. \tag{2}$$

<u>Proof</u>: It is sufficient, from formula (1), to find σ_1 and σ_2. Since G is of index μ the fundamental domain D_G consists of $2\mu = \sigma_2$ 2-simplexes in the standard triangulation. The number of 1-simplexes is simply

$$\sigma_1 = \frac{1}{2}\sum_{k=1}^{\sigma_0}(2n_k) = \sum_{k=1}^{\sigma_0} n_k \;, \tag{3}$$

i.e., the total number of 1-simplexes emanating from vertices divided by two (since each is counted twice). We can simplify (3) by breaking it up into three sums

$$\sigma_1 = \sum_{k=1}^{\lambda_i} n_k + \sum_{k=\lambda_i+1}^{\lambda_i+\lambda_\rho} n_k + \sum_{k=\lambda_i+\lambda_\rho+1}^{\sigma_0} n_k \tag{4}$$

corresponding to the points equivalent to i, ρ, and ∞. We now claim that each sum is equal to μ, i.e., $\sigma_1 = 3\mu$. We show this for the first sum only, the argument being the same for the other two. In any 2-simplex there is one vertex equivalent to i and there are two 1-simplexes of the triangle having that point as a common vertex. Each 1-simplex belongs to two 2-simplexes, and there are 2μ 2-simplexes, so that a total of 2μ 1-simplexes emanate from points equivalent to i. But since no 1-simplex connects two points equivalent to i, this number is also

$$\sum_{k=1}^{\lambda_i} (2n_k) = 2\mu .$$

Substituting $\sigma_1 = 3\mu$, $\sigma_2 = 2\mu$ in (1) proves our result.

Lemma 7: Let $G \subset \Gamma$ be a normal subgroup. Then Γ acts as a group of conformal transformations on $\overline{H/G}$ under which all points equivalent to i (respectively ρ, ∞) are equivalent.

Proof: Let $z_1 \sim_G z_2$, L in Γ; then $Lz_1 \sim_G Lz_2$. In fact, $z_1 = gz_2$ with g in G, so that $Lz_1 = (LgL^{-1})Lz_2$; and since LgL^{-1} is in G, $Lz_1 \sim_G Lz_2$. Also, Γ acts conformally on $\overline{H}$, from which it follows easily that Γ acts conformally on $\overline{H/G}$. Finally, Γ takes any point equivalent to i into any other such point, (and similarly for ρ, ∞).

It follows from Lemma 7, that all vertices of $\overline{H/G}$ equivalent to i under Γ have the same number of 1-simplexes meeting there (and similarly for ρ, ∞). Let $2n_i$, $2n_\rho$, and $2n_\infty$ be the number of 1-simplexes meeting at typical points equivalent (respectively) to i, ρ, and ∞. We call the triple (n_i, n_ρ, n_∞) the branch schema for G. Recall that we have the following conditions:

$$(5) \qquad \begin{aligned} n_i &= 1 \text{ or } 2 ; \\ n_\rho &= 1 \text{ or } 3 ; \\ n_\infty &= \text{any positive integer.} \end{aligned}$$

If we use the fact, verified during the proof of Theorem 5, that

$$\sum_{k=1}^{\lambda_i} n_i = \sum_{k=\lambda_i+1}^{\lambda_i+\lambda_\rho} n_\rho = \sum_{k=\lambda_i+\lambda_\rho+1}^{\sigma_0} n_\infty = \mu$$

we obtain

$$(6) \qquad \lambda_i n_i = \lambda_\rho n_\rho = \lambda_\infty n_\infty = \mu$$

and $$\sigma_0 = \lambda_i + \lambda_\rho + \lambda_\infty = \mu(\frac{1}{n_i} + \frac{1}{n_\rho} + \frac{1}{n_\infty}) ;$$

this proves:

Theorem 6: Let G be a normal subgroup of finite index μ in Γ. Then $\overline{H/G}$ is a Riemann surface of genus

(7) $$p = 1 + \frac{1}{2}\mu(1 - \frac{1}{n_i} - \frac{1}{n_\rho} - \frac{1}{n_\infty}) ,$$

where (n_i, n_ρ, n_∞) is the branch schema of G.

§5. Examples.

Let G be a normal subgroup of finite index μ in Γ. By Theorem 6 knowledge of the branch schema of G is sufficient to determine the genus of $\overline{H/G}$. Formula (5) shows that there are four types of branch schemata

$$(1, 1, n), (1, 3, n), (2, 1, n), \text{ and } (2, 3, n) .$$

Theorem 7: Γ has unique normal subgroups of indices $\mu = 1, 2, 3$, which we denote by Γ, G_2, G_3. Their schemata are, respectively:

(8) $$(1, 1, 1), (2, 1, 2), \text{ and } (1, 3, 3) .$$

If G is a normal subgroup which is not one of these three then the branch schema of G is $(2, 3, n)$ for some n; hence G contains neither S nor T.

Proof: (a) It is easy to see that the only branch schemata not of the form $(2, 3, n)$ must be those described in (8). For instance if the branch schema is $(2, 1, n)$, then, by Theorem 5, $p = 1 - \frac{\mu}{4n}(n + 2)$. Since all quantities involved are positive integers, we must have $p = 0$ and $n = \frac{2\mu}{4-\mu}$. Since n must be an integer, and n divides μ by formula (6), we must have $n = 2, \mu = 2$. Similar calculations show that $(1, 1, n)$ implies $n = \mu = 1$, and $(1, 3, n)$ implies $n = \mu = 3$.

(b) There is a unique subgroup of index 2, namely G_2. If G is of index 2, it is normal and $\Gamma/G \approx Z_2$ where Z_2 is the group of the integers mod 2. Any homomorphism h of Γ into Z_2 is determined by what it does to the generators T, S of Γ, and must be compatible with the relations $(TS)^3 = S^2 = I$. The only non-trivial homomorphism of Γ onto Z_2 is given by $h(T) = h(S) = 1$. The kernel is the unique subgroup of index 2. G_2 can be shown to be the subgroup generated by all squares of elements of Γ. In particular, the transformations

$$T_1: z \longrightarrow z + 2 \qquad T_1 = T^2$$
$$T_2: z \longrightarrow \frac{z - 1}{z} \qquad T_2 = TS = (ST^{-1})^2$$

belong to G_2, and can be used to construct the fundamental domain of G_2 (see Figure 3). The branch schema is (2, 1, 2).

(c) There is a unique normal subgroup of index 3, namely G_3. This is shown as above. The only homomorphisms of Γ onto Z_3 are given by

$$h(S) = 0, \qquad h(T) = 1 ;$$
$$h(S) = 0, \qquad h(T) = 2 .$$

But these two homomorphisms have the same kernel G_3, which can be shown to be the subgroup generated by all cubes of elements in Γ. The transformations:

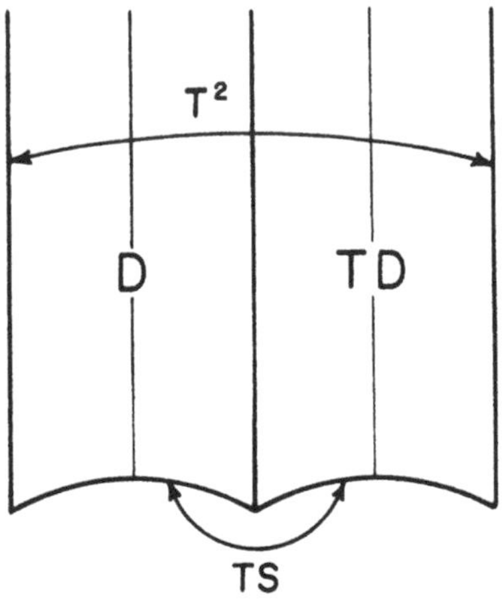

Figure 3.

$$T_1: z \longrightarrow z + 3 \qquad T_1 = T^3$$
$$T_2: z \longrightarrow -\frac{z + 2}{z + 1} \qquad T_2 = T^{-1}ST = (T^{-1}ST)^3$$
$$T_3: z \longrightarrow -\frac{1}{z} \qquad T_3 = S = S^3$$
$$T_4: z \longrightarrow \frac{z - 2}{z - 1} \qquad T_4 = TST^{-1} = (TST^{-1})^3$$

all belong to G_3 and can be used to construct its fundamental domain (see Figure 4). The branch schema is (1, 3, 3).

<u>Corollary</u>: If G is a normal subgroup of index $\mu \geq 4$ in Γ then $\mu \equiv 0 \pmod 6$, and the genus of $\overline{H/G}$ is

$$p = 1 + \frac{\mu(n_\infty - 6)}{12\, n_\infty} .$$

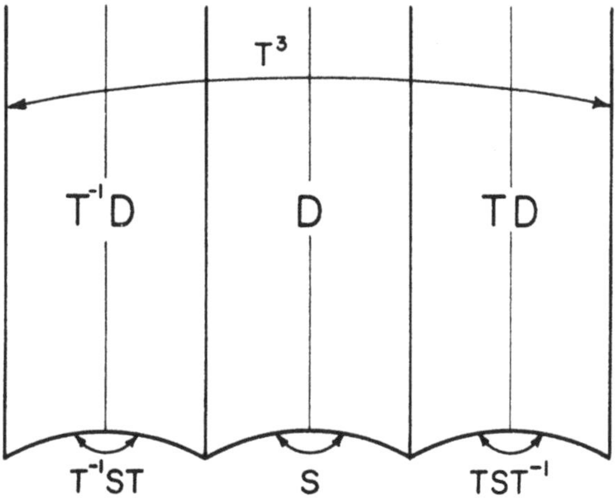

Figure 4

Theorem 8: The genus of Γ_q is:

(a) $p = 0$ if $q = 2$

(b) $p = 1 + \frac{q^2(q - 6)}{24} \prod_{p|q} (1 - \frac{1}{p^2})$ if $q > 2$.

Proof: The index of Γ_q was computed at the end of §3. It is larger than 4 whenever $q > 1$; thus the above corollary is applicable. It remains to compute n_∞. This is the number of inequivalent powers of T, which is q, a complete set of representatives being:

$$T^s = \begin{pmatrix} 1 & s \\ 0 & 1 \end{pmatrix}, \quad s = 0, 1, \ldots, q - 1.$$

We tabulate our results on Γ_q for some small values of q:

q	μ	p	q	μ	p
2	6	0	7	168	3
3	12	0	8	192	5
4	24	0	9	324	10
5	60	0	10	360	13
6	72	1	11	660	26

Remark: One can check easily that the only normal subgroups of genus zero, besides the subgroups Γ, G_2, G_3 mentioned previously, are Γ_q, for $q = 2, 3, 4$ and 5. The quotient group Γ/Γ_q acts on the quotient space $\overline{H/\Gamma_q}$, which is simply the sphere. The groups Γ/Γ_q, for $q = 2, 3, 4$, and 5, act as the dihedral group of order 6, the tetrahedral group, the octahedral group, and the icosahedral group respectively.

For future reference, we shall construct the fundamental domains of two subgroups.

(1) The first group is Γ_2. We have $\Gamma/\Gamma_2 \approx SL(2, Z_2)$, and take as coset representatives the following:

$$I = \begin{pmatrix} 1 & 0 \\ 0 & 1 \end{pmatrix} \qquad z \longrightarrow z$$

$$T = \begin{pmatrix} 1 & 1 \\ 0 & 1 \end{pmatrix} \qquad z \longrightarrow z + 1$$

$$S = \begin{pmatrix} 0 & -1 \\ 1 & 0 \end{pmatrix} \qquad z \longrightarrow -\frac{1}{z}$$

$$TS = \begin{pmatrix} 1 & -1 \\ 1 & 0 \end{pmatrix} \qquad z \longrightarrow \frac{z-1}{z}$$

$$TST = \begin{pmatrix} 1 & 0 \\ 1 & 1 \end{pmatrix} \qquad z \longrightarrow \frac{z}{z+1}$$

$$TSTS = \begin{pmatrix} 0 & 1 \\ -1 & 1 \end{pmatrix} \qquad z \longrightarrow \frac{1}{-z+1}$$

A fundamental domain for Γ_2 is given in Figure 5.

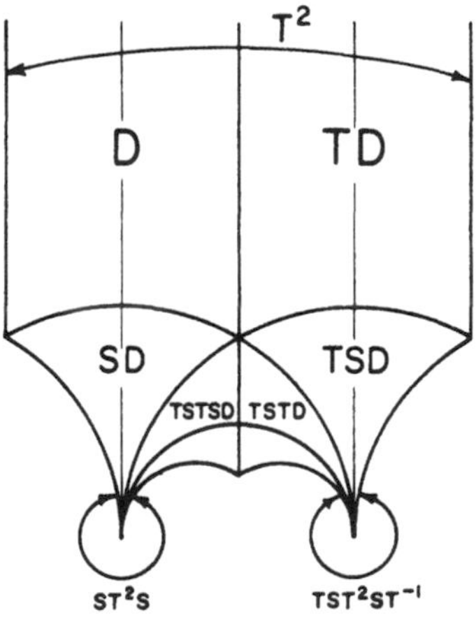

Figure 5.

(2) The group G_θ generated by T^2: $z \longrightarrow z + 2$ and S: $z \longrightarrow -\frac{1}{z}$. Two possible fundamental domains are given in Figure 6; 6b is obtained from 6a by translating the two left triangles by T^2.

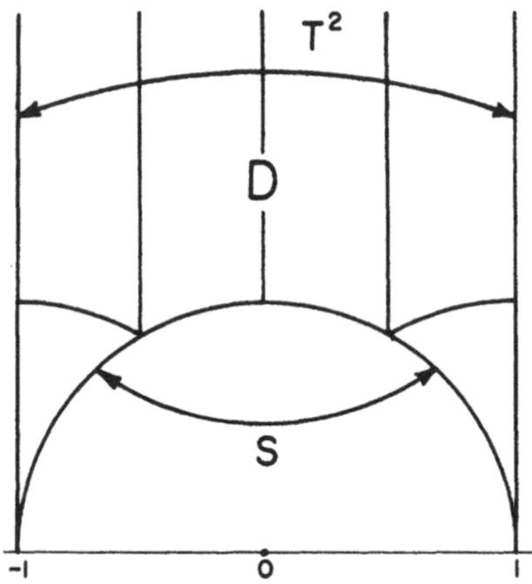

Figure 6a.

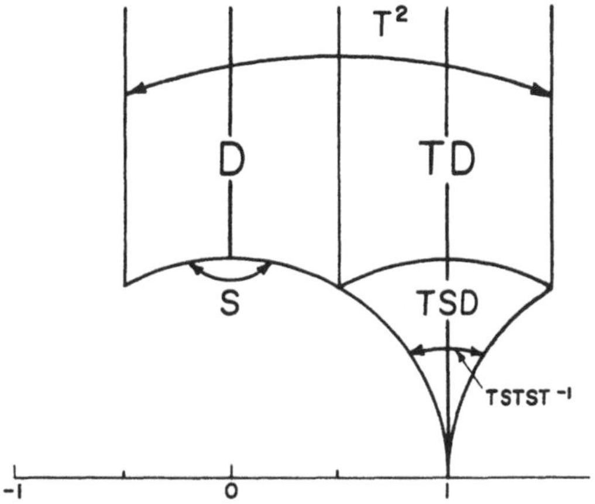

Figure 6b.

An argument similar to that of Theorem 2 shows that the transforms of the fundamental domain by elments of G_θ cover the upper half plane. So the index of G_θ is at most three, since the domain contains three copies of D, the fundamental domain for Γ. Since S is in G_θ, and S is not in G_2, G_θ is not of index 2. $G_\theta \neq \Gamma$, since T is not in G. This can be shown by proving that if W is a word in T^2 and S then $W(0)$ is in $\{\infty, \frac{2n}{2m+1}, \frac{2m+1}{2n} \mid n, m \text{ integers}\}$. Since $W(0)$ is never 1, it follows that T is not in G. $G_\theta \neq G_3$, since T^2 is in G_θ but T^2 is not in G_3. Thus G_θ is a non-normal subgroup of index 3. Its genus is easily seen to be zero. Finally one can see from the fundamental domains that $G_\theta = \Gamma_2 \cup \Gamma_2 S$, so that Γ_2 is a normal subgroup of index 2 in G_θ.

BIBLIOGRAPHY

[1]. Ford, L. R., Automorphic Function, Chelsea, (1951) [Chapter 1 of Ford's book discusses the material reviewed briefly in Section 1].

[2]. Klein, F., Vorlesungen über die Theorie der elliptischen Modulfunktionen, Leipzig, (1890).

[3]. Springer, G., Introduction to Riemann Surfaces, Addison-Wesley, (1957).

[4]. Weyl, H., Die Idee der Riemannschen Fläche, Chelsea, (1947).

CHAPTER II:

MODULAR FORMS

§6. Introduction.

It will be convenient to fix the following notation:

H: the upper half plane, $\{z \mid \operatorname{Im} z > 0\}$;

G: a subgroup of finite index μ in Γ, where Γ is the inhomogeneous modular group of linear fractional transformations;

H/G: the identification space of H modulo G, considered as a Riemann surface;

$\mathcal{S}$: the compactification of H/G obtained by adding parabolic points with the appropriate local coordinates.

It is of interest to study those meromorphic functions on H which are invariant under all transformations in the group G, that is, such that $f(Tz) = f(z)$ for all T in G. However this is too restrictive for many purposes. We shall consider instead those meromorphic functions $f(z)$ such that $f(Tz)$ and $f(z)$ have the same zeroes and poles, for all T in G. In that case, $\frac{f(Tz)}{f(z)} = \mu_T(z)$ is a holomorphic function on H which is never zero. There is moreover a consistency condition satisfied by the class of functions $\{\mu_T(z) \mid T \text{ in } G\}$: $\mu_{TS}(z) = \mu_T(Sz)\mu_S(z)$ for T,S in G. It follows from the chain rule for derivatives, and the fact that the maps are analytic automophisms, that these two requirements are satisfied by the classes:

$$\mu_T(z) = J_T(z)^k = \left(\frac{dT}{dz}\right)^k .$$

It can be shown [1] that these classes of functions are, essentially, the only ones satisfying these two conditions. Recalling that $J_T(z) = \frac{dT}{dz} = (cz + d)^{-2}$ if $T: z \longrightarrow \frac{az + b}{cz + d}$, we make the:

Definition: An unrestricted modular form of weight k for G is a meromorphic function $f(z)$ on H such that $f\left(\frac{az + b}{cz + d}\right) = (cz + d)^{2k}f(z) = J_T(z)^{-k}f(z)$ for all transformations $T: z \longrightarrow \frac{az + b}{cz + d}$ belonging to G, where k is an integer.

In order to motivate some of the subsequent definitions, we define a k-differential, dZ^k, on a Riemann surface to be a correspondence which associates to each point P of the surface and each local parameter τ at P a meromorphic function $g(\tau)$, such that $dZ^k = g(\tau)d\tau^k$; and if $t = t(\tau)$ is another local parameter at P, and $dZ^k = g_1(t)dt^k$, then $g(\tau) = g_1(t)(\frac{dt}{d\tau})^k$. (For further details see [3]). The cases $k = 0, 1$ are well known to us, being meromorphic functions and ordinary differentials on the Riemann surface. We note that if dZ^k is a k-differential, given locally by $dZ^k = g(\tau)(d\tau)^k$, and f is a meromorphic function, then $f(\tau)\, g(\tau)\, (d\tau)^k$ is also a k-differential.

Let $f(z)$ be an unrestricted modular form of weight k for G. Since $(dz)^k$ is a k-differential for H, $f(z)\,(dz)^k$ is a k-differential on H which is invariant under G in the sense that $f(Tz)\,(d(Tz))^k = f(z)\, J_T(z)^{-k}\, (J_T(z))^k\, (dz)^k = f(z)(dz)^k$, for all transformations T in G. This gives us a meromorphic k-differential on H/G.* We would like to extend it to be meromorphic on $\mathcal{S}$. This is easy at the infinite parabolic point. The local coordinate at $i\infty$ is $\zeta = e^{2\pi i\, z/q}$, q being the least positive integer such that $z \longrightarrow z + q$ is in the group G. Let $f(z) = \hat{f}(\zeta)$ (which is well defined since $f(z + q) = f(z)$); then the

---

* This is not strictly true. $f(z)(dz)^k$ will represent a k-differential at regular points. If z_0 is an elliptic point of order e, the coordinates on H/G are given by $\tau = (Sz)^e$ where z are plane coordinates and S takes $\{z_0, \bar{z}_0\}$ into $\{0, \infty\}$. We must therefore take the differential at z_0 in H/G to be $g(\tau)(d\tau)^k$, where

$$g(\tau) = \{e^{-1}\, \tau^{(1/e-1)}\, J_{S^{-1}}(\tau^{1/e})\}^k\, \hat{f}(\tau)$$

and $\hat{f}(\tau) = f(z)$, as a small calculation shows.

k-differential is locally $g(\zeta)d\zeta^k$ where $g(\zeta) = \frac{\hat{f}(\zeta)}{\zeta^k} (\frac{q}{2\pi i})^k$ and where the factor has been added to allow for the change in coordinates. It is thus natural to say that $f(z)$ is meromorphic at $i\infty$ if $\hat{f}(\zeta)$ is meromorphic at 0. The finite parabolic points, if any exist, are handled by reducing the problem to the above case. Let a/c be a finite parabolic vertex, and S be some transformation in Γ such that $S(a/c) = \infty$. Let $\mathcal{S}$ be the Riemann surface of G, and $\tilde{\mathcal{S}}$ be that of the transformed group SGS^{-1}. We have a natural conformal map $\tilde{S}: \mathcal{S} \longrightarrow \tilde{\mathcal{S}}$ defined by $S(\{x\}_G) = \{Sx\}_{SGS^{-1}}$ ($\{x\}_G$ being the equivalence class of x under G). It can be checked that $\tilde{S}$ is conformal by observing that $\tilde{S}$ takes regular, elliptic, parabolic points into points of similar type. The k-differential $f(z)(dz)^k$ on H induces under S a k-differential $g(w)dw^k = f(S^{-1}w)\, d(S^{-1}w)^k = f(S^{-1}w)\, J_{S^{-1}}(w)^k dw^k$. It is then natural to say that $f(z)$ is meromorphic at a/c if $f(w)$ is meromorphic at $i\infty$.

<u>Definition</u>: If f is an unrestricted modular form of weight k for G, the <u>S-transform of</u> f is $g(w) = J_{S^{-1}}(w)^k f(S^{-1}w)$.

<u>Lemma 1</u>: The S-transform, $g(w)$, of an unrestricted modular form of weight k for G, $f(z)$, is an unrestricted modular form of weight k for the transformed group $G^* = SGS^{-1}$.

<u>Proof</u>: Let T^* be in G^*, so that $T^* = STS^{-1}$ with T in G. Then:

$$\begin{aligned} g(T^*w) &= g(STS^{-1}w) = J_{S^{-1}}(STS^{-1}w)^k f(TS^{-1}w) \\ &= J_{S^{-1}}(STS^{-1}w)^k J_T(S^{-1}w)^{-k} f(S^{-1}w) \\ &= \left[J_{S^{-1}}(STS^{-1}w)\, J_T(S^{-1}w)^{-1}\, J_{S^{-1}}(w)^{-1}\right]^k g(w) \\ &= \left[J_{STS^{-1}}(w)\right]^{-k} g(w) \\ &= J_{T^*}(w)^{-k} g(w) . \end{aligned}$$

<u>Definition</u>: The modular form $f(z)$ is <u>holomorphic at</u> ∞ if $\hat{f}(\zeta)$ is holomorphic in $|\zeta| < 1$, where $\zeta = e^{2\pi i z/q}$ and $\hat{f}(\zeta) = f(z)$. In particular, $\hat{f}(\zeta)$ has a Taylor expansion in ζ

$$\hat{f}(\zeta) = \sum_{m=0}^{\infty} a_m \zeta^m .$$

This induces a Fourier expansion for $f(z)$

$$f(z) = \sum_{m=0}^{\infty} a_m e^{2\pi i m z/q} .$$

We call a_0 the value of $f(z)$ at ∞; if a_n is the first non-zero coefficient, n is called the order of $f(z)$ at ∞.

Definition: Let p be a parabolic fixed point of G, not ∞. Let S in Γ map p into ∞, and $g(z)$ be the S-transform of $f(z)$. Then $f(z)$ is holomorphic at p if $g(z)$ is holomorphic at ∞. The value and order of $f(z)$ at p are those of $g(z)$ at ∞.

Remark: The above definition is independent of S. In fact if S' is another transformation satisfying our hypothesis and $g'(z)$ is the S'-transform of f, then $g'(z) = g(z + c)$ for some integer c. From the fact that $g(z)$ is holomorphic at ∞ the desired result follows immediately.

Definition: A modular form is an unrestricted modular form which is holomorphic at all points of H and at all parabolic vertices of the group.

§7. A Review of Some Function Theory on Riemann Surfaces.

Let $\mathcal{S}$ be a compact Riemann surface, $f(z)$ be a meromorphic function defined on $\mathcal{S}$, and p be a point of $\mathcal{S}$. In local coordinates at p, $f(z)$ has a Laurent expansion

$$f(z) = z^n \sum_{m=0}^{\infty} a_m z^m ,$$

(with $a_0 \neq 0$). We define n to be the order of $f(z)$ at p, written $\nu_p(f) = n$. If $\omega(z)$ is a meromorphic differential form on $\mathcal{S}$, given locally by $\omega(z) = f(z)dz$, we define the order of $\omega(z)$ at p by $\nu_p(\omega) = \nu_p(f)$. Note that the above definition is independent of the parametric representation. It is useful to notice that our definitions are not vacuous, that is that there exist non-trivial meromorphic functions and differential forms on $\mathcal{S}$.

We form the free abelian group generated by the points of $\mathcal{S}$, and call it the group of divisors on $\mathcal{S}$. In particular, a divisor θ is a formal sum

$$\theta = \sum_{p \text{ in } \mathcal{S}} \mu_p\, p ,$$

where μ_p is an integer, which is zero except for a finite number of points p. Since a meromorphic function f or a meromorphic differential form ω has only a finite number of zeroes and poles, we define the divisor of f by

$$\theta(f) = \sum_{p \text{ in } \mathcal{S}} \nu_p(f)\cdot p ,$$

and the divisor of ω to be

$$\theta(\omega) = \sum_{p \text{ in } \mathcal{S}} \nu_p(\omega)\cdot p .$$

Definition: The order of the divisor

$$\theta = \sum_{p \text{ in } \mathcal{S}} \mu_p\cdot p$$

is

$$|\theta| = \sum_{p \text{ in } \mathcal{S}} \mu_p .$$

This defines a homomorphism from the group of divisors to the group of integers.

As a consequence of the residue theorem on compact Riemann surfaces the divisors of meromorphic functions are in the kernel of that homomorphism that is, $|\theta(f)| = 0$. And for any differential form ω, $|\theta(\omega)| = 2(g - 1)$, where g is the genus of $\mathcal{S}$.

We can endow the group of divisors with a partial ordering by setting $\theta \geq \theta'$ if $\mu_p \geq \mu_p'$ for all p in $\mathcal{S}$. This order is compatible with the group structure. Also if θ is positive so is $|\theta|$, though not conversely. Let f be any meromorphic function such that $\theta(f)$ is positive; then $\nu_p(f) \geq 0$ for all p in $\mathcal{S}$, which shows that f is holomorphic and hence a constant since $\mathcal{S}$ is compact.

Riemann-Roch Theorem: Let $\mathcal{M}$ be the complex vector space of meromorphic functions on $\mathcal{S}$, and for any divisor θ let $\mathcal{L}(\theta)$ be the complex vector space $\mathcal{L}(\theta) = \{f \text{ in } \mathcal{M} \mid \theta(f) + \theta \geq 0\}$. Then for any differential form ω, $\dim \mathcal{L}(\theta) = \dim \mathcal{L}(\dot{\theta}(\omega)-\theta) + |\theta| + 1 - g$, where g is the genus of $\mathcal{S}$.

For the proof of this theorem, and more detailed discussion of the concepts involved, see references [3] of Chapter I, or reference [2] at the end of this Chapter.

§8. The Dimension of the Space of Modular Forms.

The set of modular forms of weight k for a subgroup G of Γ forms a complex vector space. In this section we shall compute the dimension of this space, which will be denoted by $\delta_k(G)$.

A meromorphic function $f^*(z)$ on $\mathcal{S}$, the Riemann surface of G, induces a G-invariant meromorphic function $f(z)$ on H (together with the parabolic vertices of G) defined by $f(z) = f^*(z)$, where z is the point on $\mathcal{S}$ corresponding to the point z in H. Similarly every differential form ω^* on $\mathcal{S}$ induces a differential form ω on H.

In the last section we defined the concept of order of a function or differential form for compact Riemann surfaces; the compactness was of course not needed, so we can define the order of a meromorphic function f (differential form ω) at a point p in H also, and write $n_p(f)(n_p(\omega))$ for this order. Our first task is to compare $\nu_p(f^*)$ and $n_p(f)$, as well as $\nu_p(\omega^*)$ and $n_p(\omega)$, in the obvious notation. We must distinguish three cases:

(a) Regular points. In this case we can find a neighbourhood of p on H which is mapped conformally onto a neighbourhood of p in $\mathcal{S}$. Thus at a regular point p, $n_p(f) = \nu_p(f^*)$ and $n_p(\omega) = \nu_p(\omega^*)$.

(b) Elliptic points. Let p be an elliptic fixed point of period e. Local coordinates τ about p in $\mathcal{S}$ were obtained by first mapping a neighbourhood of p on H conformally onto a neighbourhood of zero with t and $t \exp\left(\frac{2\pi i}{e}\right)$ identified, and then taking $\tau = t^e$. Let

$$f^*(\tau) = \tau^{\nu_p(f^*)} \sum_{n=0}^{\infty} a_n \tau^n$$

with $a_o \neq o$;

then

$$f^*(t^e) = f(t) = t^{e\nu_p(f^*)}\sum_{n=0}^{\infty} a_n t^{en} = t^{n_p(f)}\sum_{n=0}^{\infty} a_n t^{en}$$

so that $n_p(f) = e\nu_p(f^*)$, and, for differential forms, $\omega^*(\tau) = f^*(\tau)d\tau = f^*(t^e)et^{e-1}dt = f(t)dt$ so that $n_p(\omega) = e\nu_p(\omega^*) + e - 1$.

(c) Parabolic points. Let p be a parabolic point which compactifies q copies of the fundamental domain D. We obtain local coordinates τ at p, on $\mathscr{S}$, by first mapping a neighbourhood of p conformally onto a neighbourhood of $i\infty$ in such a way that t and $t + q$ are identified, and then taking $\tau = e^{2\pi i\, t/q}$. Let

$$f^*(\tau) = \tau^{\nu_p(f^*)}\sum_{n=0}^{\infty} a_n \tau^n, \quad a_o \neq 0;$$

then

$$f^*(e^{2\pi i\, t/q}) = f(e^{2\pi i\, t/q}) = e^{2\pi i t\, \nu_p(f^*)/q}\sum_{n=0}^{\infty} a_n e^{2\pi i n t/q} =$$

$$= e^{2\pi i t\, n_p(f)/q}\sum_{n=0}^{\infty} a_n e^{2\pi i n t/q},$$

so that $n_p(f) = \nu_p(f^*)$, and for differential forms

$$\omega^*(\tau) = f^*(\tau)d\tau = f^*(e^{2\pi i t/q})(\tfrac{2\pi i}{q})\, e^{2\pi i t/q}\, dt = f(t)dt,$$

so that $n_p(\omega) = \nu_p(\omega^*) + 1$.

Let us now select an arbitrary differential form $\omega^*(z) = h^*(z)dz$ on $\mathscr{S}$, which will remain fixed for this discussion. This form induces a G-invariant form $\omega(z) = h(z)dz$ on H, so that $\omega(Tz) = h(Tz)\, d(Tz) = h(Tz)\, J_T(z)dz = h(z)dz$ and hence $h(Tz) = J_T(z)^{-1}\, h(z)$. Now let $f(z)$ be a holomorphic modular form of weight k; then $g(z) = \frac{f(z)}{(h(z))^k}$ is a G-invariant meromorphic function which induces a meromorphic function $g^*(z)$ on $\mathscr{S}$. The map ψ: $f(z) \longrightarrow g^*(z)$ is clearly an isomorphism between the vector space of meromorphic modular forms of weight k, and the space of meromorphic functions on $\mathscr{S}$. Thus the dimension of the space of modular forms of weight k is simply the dimension of the space of those meromorphic functions $g^*(z)$ on $\mathcal{S}$ such that $h(z)^k g(z)$ is holomorphic on H and at the parabolic vertices. With the notation used above for order on H, this condition is that $n_p(g) + kn_p(h) \geq 0$. Since we are going to solve the problem on the Riemann surface, let us rewrite that condition in terms of order on $\mathcal{S}$. Our previous calculations connecting the two concepts of order lead to the following result:

(a) <u>at a regular point</u>: $\nu_p(g^*) + k\nu_p(\omega^*) \geq 0$;

(b) <u>at an elliptic point of period e</u>;

$$\nu_p(g^*) + k\nu_p(\omega^*) + [k(1 - \tfrac{1}{e})] \geq 0 \ ;$$

(c) <u>at a parabolic point</u>: $\nu_p(g^*) + k\nu_p(\omega^*) + k \geq 0$.

Let ϵ_i be a typical elliptic point of period e_i, and let $p_1, \ldots, p_\sigma$ be the parabolic points. We define the divisor

$$\theta_o = k\theta(\omega^*) + \sum_i [k(1 - \frac{1}{e_i})]\, \epsilon_i + \sum_{i=1}^{\sigma} k \cdot p_i \quad .$$

The conditions above in the divisor notation are just that $\theta(g^*) + \theta_o \geq 0$. Thus we have $\delta_k(G) = \dim \ (\theta_o)$ in our earlier notation.

<u>Theorem 1</u>: The dimension of the space of modular forms of weight k for G is:

$$\delta_k(G) = \begin{cases} 0, & \text{if } k \leq 1 \\ 1, & \text{if } k = 0 \\ (2k - 1)(g - 1) + \sigma k + \sum_i [k(1 - \frac{1}{e_i})], & \text{if } k \geq 1, \end{cases}$$

where g is the genus of $\mathcal{S}$, σ is the number of parabolic fixed points, and the sum runs through the elliptic fixed points of G in $\mathcal{S}$ of periods e_i .

<u>Proof</u>: (1) If $k = 0$ then $\theta_o = 0$. We have seen already that $\mathcal{L}(0)$ is the space of all constants, hence has dimension 1.

(2) if $k \leq -1$ then we have for every function $g^*(z)$ in $\mathcal{L}(\theta_o)$ that

$$|\theta(g^*)| \geq - |\theta_o| = -2k(g - 1) - k\sigma - \sum_{\substack{\text{elliptic} \\ \text{vertices}}} [k(1 - \frac{1}{e_i})] \ .$$

The right hand side is strictly positive. If $g \geq 1$ this is trivial since there is at least one parabolic vertex. If the genus is zero, Theorem 5 of Chapter I shows that there must be at least three vertices (parabolic or elliptic); since $e = 2$ or 3, it then follows as desired. Since the order of a meromorphic function is zero, there can exist no non-trivial function $g^*(z)$ and $\delta k(G) = 0$ in this case.

(3) If $k \geq 1$, the Riemann-Roch Theorem gives

$$\delta_k(G) = 1 - g + |\theta_0| + \dim \mathcal{L}(\theta(\omega^*) - \theta_0) =$$
$$= (2k - 1)(g - 1) + k\sigma + \sum_i [k(1 - \frac{1}{e_i})] + \dim \mathcal{L}(\theta(\omega^*) - \theta_0).$$

A computation almost identical with that in case (2) shows that $\dim \mathcal{L}(\theta(\omega^*) - \theta_0) = 0$, from which the formula given follows.

<u>Examples</u>: Theorem 1, together with the results of Chapter I, gives, after a short calculation which we leave to the reader:

(1) For the full modular group Γ:

$$\delta_k(\Gamma) = \begin{cases} [k/6] & \text{if } k \equiv 1 \pmod 6, \\ [k/6] + 1 & \text{if } k \not\equiv 1 \pmod 6; \end{cases}$$

in particular:

$$\delta_1 = 0; \quad \delta_2 = \delta_3 = \delta_4 = \delta_5 = \delta_7 = 1; \quad \delta_6 = \delta_8 = \delta_9 = \delta_{10} = \delta_{11} = \delta_{13} = 2 .$$

(2) For the principal congruence subgroup of level 2, Γ_2:

$$\delta_k(\Gamma_2) = k + 1 .$$

(3) For the principal congruence subgroup of level $q \geq 3$, Γ_q:

$$\delta_k(\Gamma_q) = \left(\frac{(2k - 1)q + 6}{24}\right) q^2 \prod_{p|q} (1 - \frac{1}{p^2}) ;$$

for instance: $\delta_k(\Gamma_3) = 2k + 1$, $\delta_k(\Gamma_4') = 4k + 2$, $\delta_k(\Gamma_5) = 10k + 1$, $\delta_k(\Gamma_6) = 12k + 1$.

(4) For the subgroup G_θ generated by $T^2: z \longrightarrow z + 2$ and $S: z \longrightarrow -\frac{1}{z}$, we have:

$$\delta_k(G_\theta) = [k/2] .$$

BIBLIOGRAPHY

[1]. Petersson, H., Monatshefte für Mathematik, vol. 53, (1949) pp. 17-41.

Gunning, R. C., American Journal of Mathematics, vol. 78, (1956) pp. 357-383.

[2]. Schiffer, M., and Spencer, D. C , Functionals on Riemann Surfaces, Princeton University Press, (1954).

CHAPTER III:

POINCARÉ SERIES

§9. Construction of Modular Forms

Let G be a subgroup of finite index μ in the inhomogeneous modular group Γ. Theorem 1 of Chapter II guarantees the existence of modular forms of weight $k \geq 1$ for G. In this chapter we construct explicitly a family of modular forms which generate the space of all modular forms, in a sense which will be made clearer later.

Our construction of modular forms depends on a simple and beautiful idea of Poincaré, who used it in his work on automorphic functions. In order not to obscure the idea we "solve" a more general problem. Let $\{\mu_T(z) | T \text{ in } G\}$ be a collection of holomorphic, nowhere vanishing functions on H, satisfying the condition $\mu_{ST}(z) = \mu_S(Tz)\mu_T(z)$ for all S, T in G. We wish to construct a holomorphic function $f(z)$ on H such that $f(Tz) = \mu_T(z) f(z)$ for all T in G. Let $h(z)$ be any holomorphic function on H and write formally:

$$(1) \qquad f(z) = \sum_{T \text{ in } G} \frac{h(Tz)}{\mu_T(z)} \ ;$$

then

$$f(Sz) = \sum_{T \in G} \frac{h(TSz)}{\mu_T(Sz)}$$

$$= \sum_{T \in G} \frac{h(TSz)}{\mu_{TS}(z)} \mu_S(z)$$

$$= \mu_S(z) f(z).$$

If the series (1) converges absolutely uniformly on compact subsets of H, $f(z)$ is holomorphic and the formal computations become legitimate; we have then a solution to our problem.

Since there may be too many terms for which $\mu_T(z) \equiv 1$, we have little hope of convergence. The set $G_0 = \{T \in G \mid \mu_T(z) \equiv 1\}$ is a subgroup of G. Let $\mathscr{R}$ be a set of coset representatives of G mod G_0, so that

$$G = \bigcup_{T \in \mathscr{R}} G_0 T .$$

Suppose $h(z)$ is invariant under G_0, that is, that $h(Sz) = h(z)$ for all S in G_0, and define:

$$f(z) = \sum_{T \in \mathscr{R}} \frac{h(Tz)}{\mu_T(z)} . \tag{2}$$

We notice that $f(z)$ is independent of the coset representatives chosen, since, if T and T' are in the same coset mod G_0, $T' = ST$ for some $S \in G_0$ so that $h(Tz) = h(T'z)$ and $\mu_T(z) = \mu_{ST}(z) = \mu_S(Tz)\,\mu_T(z) = \mu_T(z)$. For any S in G

$$f(Sz) = \left(\sum_{T \in \mathscr{R}} \frac{h(TSz)}{\mu_{TS}(z)} \right) \mu_S(z) = f(s)\mu_S(z) ,$$

since $\mathscr{R}S$ is a set of coset representatives if $\mathscr{R}$ is. Thus, if the series (2) is absolutely and uniformly convergent on compact subsets of H, it represents a solution to our problem.

Consider in particular the class of functions

$$\mu_T(z) = J_T(z)^{-k} = (cz + d)^{2k} ,$$

where $T: z \longrightarrow \frac{az + b}{cz + d}$. The subgroup G_0 is the infinite cyclic subgroup of translations in G, generated by the least translation

$$T: z \longrightarrow z + q.$$

In terms of matrices, $G_0' = \{\pm\begin{pmatrix}1 & qr\\ 0 & 1\end{pmatrix} \big| r \in Z\}$. It is easily checked that $\begin{pmatrix}a & b\\ c & d\end{pmatrix}$ and $\begin{pmatrix}a' & b'\\ c' & d'\end{pmatrix}$ are two matrices of G' in the same coset mod G_0' if and only if $\{c,d\} = \pm\{c', d'\}$ and $\{a,b\} \equiv \pm\{a', b'\}$ mod q. Thus a set $\mathcal{R}$ of coset representatives can be obtained by taking an element T in G' for each pair $\{c,d\}$, $c \geq 0$ which is a second row of a matrix in G'. A typical function $h(z)$ invariant under G_0 is $h(z) = e^{2\pi i\nu z/q}$ ($\nu = 0,1 \ldots$); this suggests:

Definition: The Poincaré series of weight k and character ν for G is the series

$$\varphi_\nu(z) = \sum_{T\in\mathcal{R}} e^{2\pi i\nu\, T(z)/q}\, J_T(z)^k .$$

Theorem 1: The series

$$\sum_{m,n\in Z}{}' \frac{1}{|mz + n|^\ell},$$

where we have omitted the term $m = n = 0$, converges uniformly on compact subsets of H whenever $\ell > 2$.

Proof: Let z be fixed, $\text{Im } z > 0$; then $\{mz + n | m,n\in Z\}$ is the integer lattice generated by 1 and z. Let π_r be the parallelogram $\{\pm rz + n;\ nz \pm r | -r \leq n \leq r\}$ in the lattice (see Figure 7). We

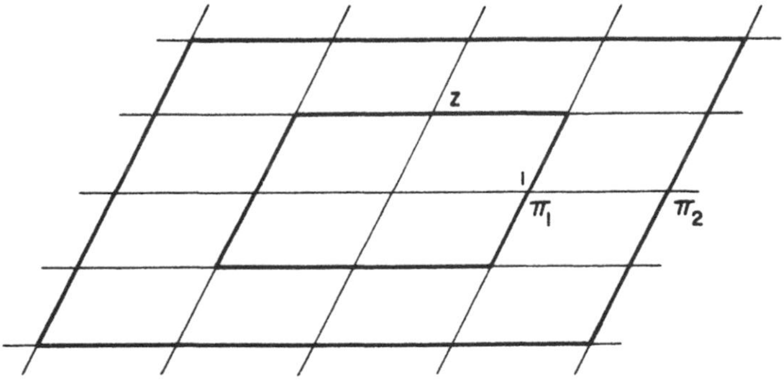

Figure 7.

sum the series over each parallelogram separately. On π_r there are $8r$ vertices. Let h be the minimum distance of π_1 to the origin; then rh is the minimum distance of π_r to the origin, so that $|mz + n| \geq rh$ if $mz + n \in \pi_r$. Thus

$$\sum_{(m,n)\in\pi_r} \frac{1}{|mz + n|^{\ell}} \leq \frac{8r}{(hr)^{\ell}} = 8h^{-\ell} \frac{1}{r^{\ell-1}}$$

and

$$\sum_{(m,n)\in\pi_r} \frac{1}{|mz + n|^{\ell}} = \sum_{r=1}^{\infty} \sum_{(m,n)\in\pi_r} \frac{1}{|mz + n|^{\ell}}$$
$$\leq 8h^{-\ell} \sum_{r=1}^{\infty} \frac{1}{r^{\ell-1}} < \infty$$

if $\ell > 2$. For uniformity of convergence, we note that our estimates depend only on h. By making h smaller, if necessary, the estimate holds uniformly for all z in any compact subset of H.

<u>Corollary</u>: The Poincare series $\varphi_\nu(z)$, for $k > 1$, $\nu \geq 0$, converges absolutely uniformly on compact subsets of H and thus represents an unrestricted modular form of weight k for G.

<u>Proof</u>: We look at a typical term, corresponding to T in $Tz = \frac{az + b}{cz + d}$. Since $\operatorname{Im} Tz = \dfrac{\operatorname{Im} z}{|cz + d|^2}$

$$\left|e^{2\pi i\nu\, T(z)/q}\,(cz + d)^{-2k}\right| = \exp\left(-\frac{2\pi\nu(\operatorname{Im} z)}{q|cz + d|^2}\right) |cz + d|^{-2k} \leq |cz + d|^{-2k}$$

for $\nu \geq 0$. Since any pair (c,d) occurs as a second row of a matrix belonging to $\mathscr{R}$ at most q times, the corollary follows immediately from the Theorem.

Our next task will be to show that the Poincare series are modular forms. To do this, we shall study their behavior at the parabolic points. We first prove a more general proposition.

Theorem 2: Let $\mathscr{P} \subset \Gamma$ be any set of transformations such that the series

$$\psi(z) = \sum_{T \in \mathscr{P}} e^{2\pi i r\, T(z)} (cz + d)^{-2k}$$

converges absolutely uniformly on compact subsets of H for all real $r \geq 0$, $k > 1$, where $Tz = \frac{az + b}{cz + d}$. Then, as $\operatorname{Im} z$ tends to $+\infty$,

$$\lim_{\operatorname{Im} z \to +\infty} \psi(z) = \begin{cases} 0 & \text{if } r > 0, \\ k & \text{if } r = 0, \text{ (} k \text{ is the number of translations in } \mathscr{P} \text{)}, \end{cases}$$

uniformly in any strip of finite width $s \leq \operatorname{Re} z \leq t$.

Proof: Write $\mathscr{P} = \mathscr{P}_1 \cup \mathscr{P}_2$, $\mathscr{P}_2$ being the set of translations in $\mathscr{R}$, $\mathscr{P}_1$ being the other transformations. In other words $\mathscr{P}_1$ consists of transformations $Tz = \frac{az + b}{cz + d}$ in $\mathscr{R}$ for which $c \neq 0$; $\mathscr{P}_2$ consists of those for which $c = 0$. We split up the sum correspondingly, $\psi = \psi_1 + \psi_2$; this is allowed by absolute convergence; and study the two series separately:

(a) We consider $\psi_2(z)$ first. If T belongs to $\mathscr{P}_2$, then for some b, $Tz = z + b$. Since $\psi_2(z)$ converges absolutely for $r = 0$:

$$\infty > \sum_{T \in \mathscr{P}_2} |cz + d|^{-2k} = \sum_{T \in \mathscr{P}_2} 1 = k\,,$$

so there are only a finite number of terms in $\mathscr{P}_2$. A typical term in $\psi_2(z)$ is

$$e^{2\pi i r(z + d)} = e^{2\pi i r(x + d)}\, e^{-2\pi y r d}\,,$$

so that

$$\lim_{\operatorname{Im} z \to \infty} \psi_2(z) = \begin{cases} k & \text{if } r = 0, \\ 0 & \text{if } r > 0, \end{cases}$$

uniformly in any finite strip $s \leq \operatorname{Re} z \leq t$.

(b) We show now that

$$\lim_{\operatorname{Im} z \to \infty} \psi_1 = 0$$

uniformly, on any finite strip $s \leq \operatorname{Re} z \leq t$. A simple calculation shows that there are constants C, ℓ, depending only on s and t, such that $|cz + d|^2 \geq \ell(c^2 + d^2)$ for $s \leq \operatorname{Re} z \leq t$ and $\operatorname{Im} z \geq C$. Thus, for all $z = x + iy$ such that $s \leq x \leq t$ and $y \geq C$:

$$(3) \qquad \left|e^{2\pi i r\, T(z)}(cz + d)^{-2k}\right| \leq \left|cz + d\right|^{-2k} \leq \frac{\ell^{-k}}{(c^2 + d^2)^k} .$$

By assumption, $\psi_2(z)$ converges for $z = i$, $r = 0$, so that

$$\sum_{T \text{ in } \mathcal{P}_1} \frac{1}{(c^2 + d^2)^k} < \infty \ ;$$

let $\mathcal{P}_1 = \mathcal{P}_3 \cup \mathcal{P}_4$ (and correspondingly $\psi_1 = \psi_3 + \psi_4$) with $\mathcal{P}_4$ finite and

$$\ell^{-k} \sum_{T \text{ in } \mathcal{P}_3} \frac{1}{(c^2 + d^2)^k} \leq \epsilon \ .$$

If $s \leq \operatorname{Re} z \leq t$, and $\operatorname{Im} z \geq C$,

$$|\psi_3(z)| \leq \ell^{-k} \sum_{T \text{ in } \mathcal{P}_3} \frac{1}{(c^2 + d^2)^k} \leq \epsilon \ ;$$

also $\psi_4(z)$ is a finite sum of terms each tending to zero uniformly as $\operatorname{Im} z$ tends to $+\infty$. Thus $\psi_1(z) = \psi_3(z) + \psi_4(z)$ tends to zero uniformly as $\operatorname{Im} z$ tends to $+\infty$, in any finite strip.

<u>Remark</u>: We have proved more than was claimed. Estimate (3) shows that $\psi(z)$ converges absolutely uniformly for $\operatorname{Im} z \geq C$ and $s \leq \operatorname{Re} z \leq t$.

<u>Theorem 3</u>: With our earlier notation, the Poincaré series

$$\varphi_\nu(z) = \sum_{T \text{ in } \mathcal{R}} e^{2\pi i\nu\, T(z)/q} \, (cz + d)^{-2k}$$

converges absolutely uniformly on compact subsets of H, for $\nu \geq 0$ and $k \geq 1$. $\varphi_\nu(z)$ converges absolutely uniformly on every fundamental domain

D for G and represents a modular form of weight k for G. Further:

(1) $\varphi_0(z)$ is zero at all finite parabolic vertices, one at $i\infty$.

(2) $\varphi_\nu(z)$ is zero at all parabolic vertices for $\nu \geq 1$.

Proof: The first claim has already been proved. Since there is only one translation in $\mathscr{H}$, our statement about the behaviour at $i\infty$ follows from Theorem 2. $\varphi_\nu(z)$ is holomorphic at $i\infty$ (as a modular form) by Riemann's theorem on removable singularities. Let p be a finite parabolic vertex of G, which is not equivalent to $i\infty$. Let $S(p) = \infty$, and $\varphi_\nu^*(z)$ be the S-transform of $\varphi_\nu(z)$:

$$\begin{aligned}\varphi_\nu^*(z) &= J_{S^{-1}}(z)^k\, \varphi_\nu(S^{-1}z)\\ &= J_{S^{-1}}(z)^k \sum_{T \text{ in } \mathscr{H}} e^{2\pi i\nu\, TS^{-1}z/q}\, J_T(S^{-1}z)^k\\ &= \sum_{T \text{ in } \mathscr{H}} e^{2\pi i\nu\, TS^{-1}z/q}\, J_{TS^{-1}}(z)^k\\ &= \sum_{T \text{ in } \mathscr{H}S^{-1}} e^{2\pi i\nu\, Tz/q}\, J_T(z)^k\end{aligned}$$

The series still satisfies the hypothesis of Theorem 2. There are no translations in $\mathscr{H}S^{-1}$, for otherwise p would be equivalent to $i\infty$. Thus, since the behaviour of φ_ν^* at $i\infty$ determines that of φ_ν at p, we obtain the desired result. The convergence on the fundamental domain follows from the remark after Theorem 2 and the corresponding statement about φ_ν^* above.

§10. The Petersson Inner Product

Definition: A cusp form of weight k for G is a modular form of weight k for G which vanishes at all parabolic vertices (cusps).

We have shown in the last section that $\varphi_\nu(z)$, $\nu \geq 1$, are cusp forms. The next section shows that the Poincaré series for $\nu \geq 1$ generate the space of cusp forms. In order to do this, we introduce here an inner product on the space of cusp forms. We are following a technique of Petersson, [1].

We denote by $d\mu(z) = dx \wedge dy = \frac{i}{2}(dz \wedge \overline{dz})$ the standard plane measure. Let f,g be two modular forms of weight $k > 0$ for G. We define a measure on H by

$$\{f,g\}(z) = f(z)\,\overline{g(z)}\,(\operatorname{Im} z)^{2(k-1)} d\mu(z) \,.$$

This is a two-form on H which satisfies the following properties:

(1) $\{f,g\} = \overline{\{g,f\}}$;

(2) $\{f,f\} \geq 0$; $\{f,f\} = 0 \iff f = 0$;

(3) $\{f,g\}(Tz) = \{f,g\}(z)$, for any T in G.

Property (3) follows from the fact that $\operatorname{Im} Tz = |J_T(z)|\operatorname{Im} z$ and $d\mu(Tz) = |J_T(z)|^2 d\mu(z)$. Let D be a fundamental domain for G; we define

$$(4) \qquad (f,g) = \int_D \{f,g\}(z) = \int_D f(z)\,\overline{g(z)}\, y^{2(k-1)} dx \wedge dy.$$

Assuming that this integral converges, it is independent of the fundamental domain D chosen, by property (3) of the two-form $\{f,g\}$. It represents a complex valued bilinear map which satisfies the properties of an inner product:

(1) $(f,g) = \overline{(g,f)}$;

(2) $(f,f) \geq 0$; $(f,f) = \iff f = 0$.

<u>Theorem 4</u>: The space of cusp forms of weight k for G is a finite dimensional Hilbert space with the Petersson inner product (4).

<u>Proof</u>: It is clearly sufficient to show that the integral (4) converges, the other statements having been proved previously. Let $Tz = z + q$ be the least translation in G. We may choose the fundamental domain D in the strip $0 \leq \operatorname{Re} z \leq q$. We decompose D into two pieces, one of infinite area containing the infinite parabolic vertex, the other containing all finite parabolic vertices: $D = D' \cup D''$,
$D' = \{z \text{ in } H | 0 \leq \operatorname{Re} z \leq q, \operatorname{Im} z \geq 1\}$ and $D'' \subseteq \{z \text{ in } H | 0 \leq \operatorname{Re} z \leq q, \operatorname{Im} z \leq 1\}$.
Since f and g are holomorphic and vanish at $i\infty$, we can expand them in Fourier series which converge absolutely uniformly for $\operatorname{Im} z \geq 1$:

$$f(z) = \sum_{\lambda=1}^{\infty} a_\lambda \, e^{2\pi i \lambda z/q} ;$$

$$g(z) = \sum_{\lambda=1}^{\infty} b_\lambda \, e^{2\pi i\lambda z/q} .$$

Thus:

$$\left| f(z)\,\overline{g(z)}\, y^{2(k-1)} \right| \le \sum_{\lambda=1}^{\infty} \sum_{\mu=1}^{\infty} \left| a_\lambda \bar{b}_\mu e^{2\pi i\lambda z/q} \, e^{-2\pi i\mu\bar{z}/q} \, y^{2(k-1)} \right|$$

(5)
$$\le \sum_{\lambda=1}^{\infty} \sum_{\mu=1}^{\infty} \left| a_\lambda b_\mu \right| e^{-2\pi y(\lambda+\mu)/q} \, y^{2(k-1)}$$

$$= \sum_{m=2}^{\infty} c_m \, e^{-2\pi my/q} \, y^{2(k-1)} \qquad (c_m \ge 0) ,$$

the last series converging absolutely uniformly for $y \ge 1$. Since the right-hand side is independent of x

$$\left| \int_{D'} f(z)\, g(z)\, y^{2(k-1)} \, dxdy \right| \le \int_{D'} \sum_{m=2}^{\infty} c_m \, e^{-2\pi my/q} \, y^{2(k-1)} \, dxdy$$

$$= q \int_{y=1}^{\infty} dy \sum_{m=2}^{\infty} c_m \, e^{-2\pi my/q} \, y^{2(k-1)} .$$

Now for any $m \ge 1$,

$$\int_{y=1}^{\infty} e^{-2\pi my/q} \, y^{2(k-1)} \, dy = e^{-2\pi m/q} \int_{1}^{\infty} e^{-2\pi m(y-1)} \, y^{2(k-1)} \, dy$$

$$\le e^{-2\pi m/q} \int_{1}^{\infty} e^{-2\pi(y-1)} \, y^{2(k-1)} \, dy$$

$$= Ke^{-2\pi m/q} .$$

Thus the integral is dominated term by term by

$$q \sum_{m=2}^{\infty} c_m K e^{-2\pi m/q} ,$$

which converges, since it is essentially equation (5) with $y = 1$. Ignoring small neighbourhoods of the finite parabolic vertices, the rest of D'' is compact and so does not raise any problems of convergence. Since there are only a finite number of parabolic vertices, we can restrict our attention to a neighbourhood N of a typical finite parabolic vertex p. Let $Sz = w$, with $S(p) = \infty$, and $f_o(w)$, $g_o(w)$ be the S-transforms of $f(z)$ and $g(z)$:

$$f(S^{-1}w) = J_{S^{-1}}(w)^{-2k} f_o(w) ,$$

$$g(S^{-1}w) = J_{S^{-1}}(w)^{-2k} g_o(w) .$$

We have

$$\int_N f(z)\, \overline{g(z)} (\operatorname{Im} z)^{2(k-1)} \tfrac{i}{2}\, dz \wedge \overline{dz} =$$

$$= \int_{SN} f_o(w)\, \overline{g_o(w)} (\operatorname{Im} w)^{2(k-1)} \tfrac{i}{2}\, dw \wedge \overline{dw} ,$$

which is precisely the integral we considered above. Since $f(z)$ and $g(z)$ vanish at p, $f_o(w)$ and $g_o(w)$ vanish at $i\infty$, which ensures the convergence of the integral.

§11. Completeness of Poincaré Series.

Theorem 5: Let $f(z)$ be a cusp form of weight $k > 1$ for G, $\varphi_\nu(z)$ be a corresponding Poincaré series with $\nu \geq 1$; then

$$(f,\varphi_\nu) = \int_D f(z)\, \overline{\varphi_\nu(z)}\, y^{2(k-1)}\, dxdy$$

(6)

$$= \frac{q^{2k}(2k-2)!}{(4\pi)^{2k-1}}\, \nu^{1-2k}\, a_\nu ,$$

where a_ν is the νth Fourier coefficient in the expansion

$$f(z) = \sum_{n=1}^{\infty} a_n e^{2\pi i n z/q}$$

Proof: Theorem 4 shows that the integral converges, since both φ_ν and f are cusp forms; and

$$(f,\varphi_\nu) = \int_D \sum_{T \text{ in } \mathcal{R}} f(z) e^{-2\pi i \nu \overline{T(z)}/q} (\overline{cz+d})^{-2k} y^{2(k-1)} \, dxdy =$$

$$= \sum_{T \text{ in } \mathcal{R}} \int_D f(z) e^{-2\pi i \nu \overline{T(z)}/q} (\overline{cz+d})^{-2k} y^{2(k-1)} \, dxdy .$$

We are justified in interchanging the series and integral, since the series converges absolutely uniformly in D and the sum is integrable. In a typical term

$$\int_D f(z) e^{-2\pi i \nu \overline{T(z)}/q} (cz+d)^{-2k} y^{2(k-1)} \, dxdy$$

substitute $w = Tz = u + iv$ to obtain

$$\int_{TD} f(w) e^{-2\pi i \nu \overline{w}/q} v^{2(k-1)} \, dudv .$$

Thus

$$(f,\varphi_\nu) = \int_{\substack{\cup (TD) \\ T \text{ in } \mathcal{R}}} f(w) e^{-2\pi i \nu \overline{w}/q} v^{2(k-1)} \, dudv,$$

since $T_1(D) \cap T_2(D)$ is a set of measure zero if $T_1 \neq T_2$. Since $H = GD = G_0 \mathcal{R} D$, it follows that $\mathcal{R} D = \bigcup_{T \text{ in } \mathcal{R}} TD$ is a fundamental domain for G_0, the group of translations generated by $Tz = z + q$. We have shown previously that $\varphi_\nu(z)$ is independent of the set $\mathcal{R}$ of coset representatives of $G \bmod G_0$. Thus we can assume that $\mathcal{R}$ was chosen in such a way that

$$\bigcup_{T \text{ in } \mathcal{R}} TD = \{z \text{ in } H \mid 0 \le \operatorname{Re} z \le q\} ,$$

he simplest fundamental domain for G_o. We have obtained

$$(f,\varphi_\nu) = \int_{x=0}^{q} \int_{y=0}^{\infty} f(z) e^{-2\pi i\nu z/q} \, y^{2(k-1)} \, dxdy \, .$$

ere $f(z)$, being a cusp form, has a Fourier expansion

$$f(z) = \sum_{\mu=1}^{\infty} a_\mu e^{2\pi i\mu z/q}$$

hich converges absolutely uniformly for $y \geq \gamma > 0$. This allows us to nterchange the summation and integration to secure

$$(f,\varphi_\nu) = \int_{x=0}^{q} \int_{y=0}^{\infty} \sum_{\mu=1}^{\infty} a_\mu e^{2\pi i\mu z/q} \, e^{-2\pi i\nu \bar{z}/q} \, y^{2(k-1)} \, dxdy$$

$$= \sum_{\mu=1}^{\infty} a_\mu \int_{x=0}^{q} e^{2\pi i(\mu-\nu)x/q} \, dx \int_{y=0}^{\infty} e^{-2\pi(\mu+\nu)y/q} \, y^{2(k-1)} \, dy$$

$$= qa_\nu \int_{y=0}^{\infty} e^{-4\pi\nu y/q} \, y^{2(k-1)} \, dy = qa_\nu \left(\frac{q}{4\pi\nu} \right)^{2k-1} \int_{0}^{\infty} e^{-t} \, t^{2k-2} \, dt.$$

Theorem 6: Every cusp form is a linear combination of the Poincaré eries $\varphi_\nu(z)$, $\nu \geq 1$.

Proof: The set of all cusp forms is a finite dimensional Hilbert pace with the Petersson inner product. The set of Poincaré series, $\nu \geq 1$, enerates a linear subspace which is necessarily closed. Any cusp form rthogonal to that subspace must have all its Fourier coefficients equal to ero by Theorem 5, thus must vanish identically.

It would be of interest to know explicitly a basis for the space of usp forms, and the linear relations satisfied by the Poincaré series. Un- ortunately, little is known about these questions, even in the simplest ases (e.g., when is $\varphi_\nu(z) = 0$?).

§12. The Fourier Coefficients of Cusp Forms

We have seen that $\varphi_0(z)$ is a modular form which is one at $i\infty$ and vanishes at the other parabolic vertices. To obtain a modular form $\varphi_0^p(z)$ which is one at the parabolic point p and which vanishes at all other parabolic points, we use the now familiar trick of sending p into $i\infty$ by means of some transformation S in Γ. Let $\varphi_0^*(z)$ be the Poincaré series for SGS^{-1} which is one at $i\infty$. Then the S^{-1}-transform of $\varphi_0^*(z)$ is the desired function $\varphi_0^p(z)$.

Consider an arbitrary modular form $f(z)$ with value a_p at the parabolic vertex p, and let $\varphi_0^p(z)$ be the modular form described above. Then

$$f(z) - \sum_{\substack{\text{parabolic} \\ \text{points } p}} a_p \, \varphi_0^p(z) = g(z)$$

is a cusp form. We shall compute the Fourier coefficients of $\varphi_0^p(z)$ for Γ_q in the next chapter. We devote this section to the Fourier coefficients of cusp forms [2].

Let $\varphi_\nu(z)$ be the Poincaré series for G,

$$\varphi_\nu(z) = \sum_{T \text{ in } \mathscr{R}} e^{2\pi i\nu\, T(z)/q} \, (cz + d)^{-2k} ,$$

where $\mathscr{R}$ is a complete set of coset representatives for G mod G_0, the subgroup of translations. Then

$$|\varphi_\nu(z)| \leq \sum_{T \text{ in } \mathscr{R}} e^{-2\pi\nu |cz+d|^2 \, y/q} \; \frac{1}{|cz + d|^{2k}} \; .$$

The right-hand side depends only on the second rows of the transformations of $\mathscr{R}$; it follows from our comments before the definition of the Poincaré series that there are at most q transformations $\begin{pmatrix} a & b \\ c & d \end{pmatrix}$ in $\mathscr{R}$ with a given second row (c,d). If $\mathfrak{S}$ is a complete set of coset representatives of Γ modulo the translations in Γ, then

$$|\varphi_\nu(z)| \leq q \sum_{(c,d) \text{ in } \Xi} e^{-2\pi\nu|cz+d|^{-2} y/q} \frac{1}{|cz+d|^{2k}} = q\,\psi(z).$$

In the sum representing $\psi(z)$, there is a term $c = 0$, $d = 1$; the other terms correspond to $c > 0$, $(c,d) = 1$. Thus

$$\psi(z) \leq e^{-2\pi\nu y/q} + \sum_{c=1}^{\infty} \sum_{d=-\infty}^{\infty} \frac{1}{|cz+d|^{2k}} .$$

Let

$$\psi_1(z) = \sum_{c=1}^{\infty} \sum_{|cx+d| \leq cy} \frac{1}{((cx+d)^2 + (cy)^2)^k} ,$$

$$\psi_2(z) = \sum_{c=1}^{\infty} \sum_{|cx+d| > cy} \frac{1}{((cx+d)^2 + (cy)^2)^k} .$$

There are at most $2(cy) + 1$ terms in the inner sum for $\psi_1(z)$, and therefore less than $3cy$ for y large enough; thus:

$$\psi_1(z) \leq \sum_{c=1}^{\infty} \frac{3cy}{(cy)^{2k}} = y^{1-2k} \left(\sum_{c=1}^{\infty} \frac{3}{c^{2k-1}} \right) .$$

Similarly

$$\psi_2(z) \leq \sum_{c=1}^{\infty} \sum_{(cx+d) > cy} \frac{1}{(cx+d)^{2k}} \leq \sum_{c=1}^{\infty} \int_{cy}^{\infty} \frac{dt}{t^{2k}} =$$

$$= \sum_{c=1}^{\infty} \frac{1}{(2k-1)(cy)^{2k-1}} \leq K_2 y^{1-2k} .$$

Finally, since $e^{-2\pi\nu y/q}$ tends to zero faster than y^{1-2k} as y tends to $+\infty$, for $\nu \geq 1$, we conclude that

(7) $$\psi(z) = O(y^{1-2k}) \quad \text{as} \quad y \longrightarrow +\infty ,$$

in the Hardy-Littlewood notation. Let

$$\psi_0(z) = y^k \psi(z) = \sum_{T \text{ in } \mathfrak{Z}} e^{(-2\pi\nu/q)\mathrm{Im}(Tz)} (\mathrm{Im}(Tz))^k .$$

$(\mathrm{Im}\ Tz = \frac{y}{|cz+d|^2})$. We note that $\psi_0(z)$ is independent of the choice of coset representatives for Γ modulo the subgroup of translations. From this we see that $\psi_0(z)$ is invariant under Γ. We have $\psi_0(z) = O(y^{1-k})$ as $y \longrightarrow \infty$. Thus, since $k > 1$, $\psi_0(z)$ is bounded in the fundamental domain for Γ (see Figure 1). Since $\psi_0(z)$ is invariant under Γ, it is bounded on the whole upper half plane H; $|\psi_0(z)| < K$ for all $y > 0$. Therefore

$$(8) \qquad \psi(z) = y^{-k} \psi_0(z) = O(y^{-k}) \quad \text{as} \quad y \longrightarrow 0+ .$$

Since $|\varphi_\nu(z)| \leq q\, \psi(z)$ and the cusp forms are generated by the Poincaré series $\varphi_\nu(z)$, $\nu \geq 1$, we have proved:

<u>Theorem 7</u>: Let $f(z)$ be any cusp form of weight $k > 1$ for G; then:

$$f(z) = O(y^{-k}) \quad \text{as} \quad y \longrightarrow 0+ ,$$
$$f(z) = O(y^{1-2k}) \quad \text{as} \quad y \longrightarrow +\infty .$$

<u>Theorem 8</u>: Let $f(z)$ be a cusp form with Fourier expansion

$$f(z) = \sum_{\nu=1}^{\infty} a_\nu\, e^{2\pi i \nu z/q} ;$$

The Fourier coefficients satisfy the condition

$$a_\nu = O(\nu^k) \quad \text{as} \quad \nu \longrightarrow \infty .$$

<u>Proof</u>: Since the series is absolutely uniformly convergent in any half plane contained in H we can interchange the summation and integration to secure:

$$\frac{1}{q}\int_{x=0}^{q} e^{-2\pi i \nu z/q} f(z)\, dx = \frac{1}{q}\int_{x=0}^{q} dx\, e^{-2\pi i \nu z/q} \sum_{\mu=1}^{\infty} a_\mu\, e^{2\pi i \mu z/q} =$$

$$= \frac{1}{q} \sum_{\mu=1}^{\infty} a_\mu \left(\int_{x=0}^{q} e^{2\pi i \frac{x}{q}(\mu-\nu)}\, dx \right) e^{-2\pi y(\mu-\nu)/q} = a_\nu \quad .$$

Taking absolute values of both sides and applying Theorem 7 for small y, we obtain

$$|a_\nu| \le \frac{1}{q} \int_{x=0}^{q} e^{2\pi\nu y/q}\, Ky^{-k}\, dx = Ky^{-k}\, e^{2\pi\nu y/q} \quad .$$

Let ν be sufficiently large, and set $y = \frac{1}{\nu}$; then $|a_\nu| < Ke^{2\pi/q}\, \nu^k$.

BIBLIOGRAPHY

[1]. Petersson, H., Jahresbericht der Deutschen Math. Verernigung, vol. 49 (1939), pp. 49-75.

Petersson, H., Math Ann., vol. 117 (1940), pp. 453-537.

Petersson, H., Hamb. Abh., vol. 14 (1941), pp. 22-60.

[2]. Hecke, E., Hamb. Abh., vol. 5 (1926), pp. 199-224.

For better estimates in some special cases see also:

Kloostermann, H. D., Hamb. Abh., vol. 5 (1927), pp. 337-352.

Rankin, R. A., Proc. Cambridge Philosophical Society, vol. 35 (1939), pp. 357-372.

CHAPTER IV:

EISENSTEIN SERIES

§13. Construction of the Eisenstein Series.

In this chapter we restrict ourselves to the principal congruence subgroups Γ_q, $(q = 1,2,3,\ldots)$; recall that Γ_q consists of the transformations represented by the matrix group

$$\Gamma'_q = \left\{ \begin{pmatrix} a & b \\ c & d \end{pmatrix} \middle| ad - bc = 1, \quad a \equiv d \equiv 0 \pmod q, \quad b \equiv c \equiv 0 \pmod q \right\} .$$

The subgroup of translations in Γ_q is represented by the matrices $\begin{pmatrix} 1 & rq \\ 0 & 1 \end{pmatrix}$ $r = 0, \pm 1, \pm 2, \ldots$. A complete set of coset representatives for Γ_q modulo the subgroup of translations can be obtained by selecting a transformation T in Γ_q corresponding to each pair $\{c,d\}$ such that $(c,d) = 1$* and $c \equiv 0$, $d \equiv 1 \pmod q$. (In case $q = 1,2$ we take $d \geq 0$ only since in that case both $\pm I$ belong to Γ'_q).

The Poincaré series of weight $k > 1$ and character $\nu = 0$ for Γ_q is

$$(1) \qquad \varphi_0(z) = \sum_{\substack{\{c,d\} \equiv \{0,1\} (q) \\ (c,d) = 1}} \frac{1}{(cz + d)^{2k}} .$$

This represents a modular form of weight k for Γ_q which has the value one at $i\infty$ and vanishes at all other parabolic cusps.

Let p be a finite parabolic vertex and S be a transformation in Γ such that $S(p) = i\infty$. The S^{-1} transform of $\varphi_0(z)$, $\varphi_0^*(z) = J_S(z)^k \varphi_0(Sz)$,

---

* Here (c,d) is the usual notation for the greatest common divisor of the integers c,d.

is a modular form of weight k for $S^{-1}\Gamma_q S = \Gamma_q$ which is one at p and vanishes at the other parabolic vertices; it has the series expansion

$$\varphi_o^*(z) = J_S(z)^k \,\varphi_o(Sz)$$

$$= J_S(z)^k \sum_{T \text{ in } \mathcal{T}} J_T(Sz)^k$$

$$= \sum_{T \text{ in } \mathcal{T}} (J_{TS}(z))^k$$

$$= \sum_{T \text{ in } \mathcal{T}S} (J_T(z))^k \quad .$$

Let $S = \begin{pmatrix} \alpha & \beta \\ \gamma & \delta \end{pmatrix}$, $\alpha\delta - \beta\gamma = 1$, so that $S^{-1} = \begin{pmatrix} \delta & -\beta \\ -\gamma & \alpha \end{pmatrix}$ and $S^{-1}(\infty) = p = -\frac{\delta}{\gamma}$. We check easily that $\mathcal{T}S$ contains one transformation in Γ_q corresponding to each pair $\{c,d\}$ such that $(c,d) = 1$ and $c \equiv \gamma$, $d \equiv \delta \pmod q$. Thus

(2) $$\varphi_o^*(z) = \sum_{\substack{\{c,d\} \equiv \{\gamma,\delta\}(q) \\ (c,d)=1}} (cz+d)^{-2k} \quad .$$

<u>Definition</u>: A restricted <u>Eisenstein Series</u> of weight $k > 1$ for Γ_q is a series

$$G_k^*(z;\gamma,\delta;q) = \sum_{\substack{\{c,d\} \equiv \{\gamma,\delta\}(q) \\ (c,d) = 1}} (cz+d)^{-2k} \quad .$$

Of course, this restricted Eisenstein series is a modular form of weight k for the group Γ_q .

<u>Lemma 1</u>: $G_k^*(z;\gamma,\delta;q) = G_k^*(z;\gamma',\delta';q)$ if $\{\gamma,\delta\} \equiv \{\gamma',\delta'\} \pmod q$ or if $\{\gamma,\delta\} \equiv \{-\gamma,-\delta\} \pmod q$. Except for these relations the restricted Eisenstein series are linearly independent.

<u>Proof</u>: The first statement is obvious. We have seen that there are $\frac{1}{2} q^2 \prod_{p|q} (1 - \frac{1}{p^2})$ parabolic vertices for Γ_q if $q > 2$ (see §3,4).

Since to each parabolic cusp corresponds a restricted Eisenstein series, and these must be linearly independent, there are at least $\frac{1}{2} q^2 \prod_{p|q} (1 - \frac{1}{p^2})$ linearly independent such series. Let us now count the number of distinct Eisenstein series modulo the relations noted in the lemma. The series $G_k^*(z;\gamma,\delta;q)$ is only defined for $(\gamma,\delta,q) = 1$, that is, for $\{\gamma,\delta\}$ a primitive pair mod q (see §3). We have already seen that there are $q^2 \prod_{p|q} (1 - \frac{1}{p^2})$ incongruent primitive pairs modulo q. Since $\{\gamma,\delta\}$ and $\{-\gamma,-\delta\}$ give rise to the same Eisenstein series, and $q > 2$, there are $\frac{1}{2} q^2 \prod_{p|q} (1 - \frac{1}{p^2})$ distinct Eisenstein series, which must be linearly independent. For $q = 1$ or 2, the result is also easily checked by making the obvious modifications in the proof.

<u>Definition</u>: A <u>general Eisenstein series</u> of weight k for Γ_q is a series

$$G_k(z;\gamma,\delta;q) = \sum_{\substack{\{c,d\} \equiv \{\gamma,\delta\} \\ \{c,d\} \neq \{0,0\}}} (cz + d)^{-2k} .$$

Theorem 1 of Chapter III shows that the series converges absolutely uniformly on compact subsets of H if $k > 1$, and thus represents a holomorphic function on H. The general Eisenstein series are modular functions for Γ_q, as we shall soon see.

<u>Definition</u>: $G_k(z;\gamma,\delta;q)$ is called a <u>primitive Eisenstein series</u> if $(\gamma,\delta,q) = 1$; otherwise it is <u>imprimitive</u>.

We first prove a lemma the object of which is to show that it is sufficient to study primitive Eisenstein series.

<u>Lemma 2</u>. If $(\gamma,\delta,q) = r$, then $G_k(z;\gamma,\delta;q) = \frac{1}{r^{2k}} G_k(z;\frac{\gamma}{r}, \frac{\delta}{r}; \frac{q}{r})$. The Eisenstein series on the right is clearly primitive.

<u>Proof</u>: $$G_k(z;\gamma,\delta;q) = \sum_{\{c,d\} \equiv \{\gamma,\delta\} \bmod q} \frac{1}{(cz + d)^{2k}} .$$

Since $c \equiv \gamma \equiv \gamma' r \pmod{q'r = q}$, we must have $c \equiv c'r$, with $c' \equiv \frac{\gamma}{r} \pmod{\frac{q}{r}}$; and similarly $d = d'r$, with $d' \equiv \frac{\delta}{r} \pmod{\frac{q}{r}}$. Substitution leads to the desired result.

We devote the remainder of this section to showing that the restricted Eisenstein series and the primitive general Eisenstein series span the same vector space. We conclude from this that the general Eisenstein series are modular forms for Γ_q of weight k.

Let $G_k(z;\gamma,\delta;q)$ be a primitive Eisenstein series so that $((\gamma,\delta),q) = 1$ and

$$G_k(z;\gamma,\delta;q) = \sum_{\{c,d\}\equiv\{\gamma,\delta\}(q)} (cz+d)^{-2k}$$

$$= \sum_{\substack{a=1\\(a,q)=1}}^{\infty} \sum_{\substack{\{c,d\}\equiv\{\gamma,\delta\}(q)\\(c,d)=a}} (cz+d)^{-2k}$$

Since $(a,q) = 1$, there is an integer $\bar{a}$, $1 \le \bar{a} \le q-1$, such that $a\bar{a} \equiv 1 \pmod q$; so writing $c = ac'$, $d = ad'$, in the inner sum, we obtain

$$G_k(z;\gamma,\delta;q) = \sum_{\substack{a=1\\(a,q)=1}}^{\infty} \sum_{\substack{\{c'd'\}\equiv\{\bar{a}\gamma,\bar{a}\delta\}(q)\\(c',d')=1}} (c'z+d')^{-2k}\, a^{-2k}$$

$$= \sum_{\substack{a=1\\(a,q)=1}}^{\infty} \frac{1}{a^{2k}}\, G_k^*(z;\bar{a}\gamma,\bar{a}\delta;q) \quad .$$

As a runs through the integers prime to q, $\bar{a}$ runs through the integers in the interval $1 \le x \le q-1$ which are prime to q, so we can rewrite the above expression in the somewhat neater form

$$(3) \qquad G_k(z;\gamma,\delta;q) = \sum_{\substack{a=1\\(a,q)=1}}^{q} \left(\sum_{\substack{n=1\\n\equiv 1(q)}}^{\infty} \frac{1}{n^{2k}} \right) G_k^*(z;a\gamma,a\delta;q) \quad .$$

Let $\mu(n)$ denote the Möbius function,

$$\mu(n) = \begin{cases} 1 & \text{if } n = 1 \\ 0 & \text{if } n \text{ is divisible by a square} \\ (-1)^r & \text{if } n = p_1 \cdots p_r \text{ is a factorization into distinct primes.} \end{cases}$$

We recall the elementary lemma

$$\sum_{d \mid n} \mu(d) = \begin{cases} 1 & \text{if } n = 1 , \\ 0 & \text{if } n > 1 . \end{cases}$$

Thus

$$G_k^*(z;\gamma;\delta,q) = \sum_{\substack{\{c,d\} \equiv \{\gamma,\delta\}(q) \\ (c,d)=1}}{}' (cz + d)^{-2k}$$

$$= \sum_{\{c,d\} \equiv \{\gamma,\delta\}(q)} \sum_{a \mid (c,d)} \mu(a) \; (ca + d)^{-2k}$$

$$= \sum_{\substack{a=1 \\ (a,q)=1}}^{\infty} \sum_{\substack{\{c,d\} \equiv \{\gamma,\delta\}(q) \\ a \mid (c,d)}}{}' \mu(a) \; (cz + d)^{-2k} .$$

(We are restricting a to be prime to q, since otherwise $(\gamma,\delta,q) > 1$, which is impossible for a restricted Eisenstein series.)

As before, write $c = ac'$, $d = ad'$, and let $\bar{a}$ be an integer $1 \le \bar{a} \le q$ so that $a\bar{a} \equiv 1(q)$, which is possible since a is prime to q. Then

$$G_k^*(z;\gamma,\delta;q) = \sum_{\substack{a=1 \\ (a,q)=1}}^{\infty} \sum_{\{c',d'\} \equiv \{a\gamma,a\delta\}(q)}{}' \mu(a) a^{-2k} \; (c'z + d')^{-2k}$$

$$= \sum_{\substack{a=1 \\ (a,q)=1}}^{\infty} \frac{\mu(a)}{a^{2k}} G_k(z;\bar{a}\gamma;\bar{a}\delta,q) \quad ,$$

or

(4)

$$G_k^*(z;\gamma,\delta;q) = \sum_{\substack{a=1 \\ (a,q)=1}}^{q} \left(\sum_{\substack{n=1 \\ na \equiv 1(q)}}^{\infty} \frac{\mu(n)}{n^2} \right) G_k(z;a\gamma,a\delta;q) \quad .$$

Remarks:

(1) It would have been possible to obtain equation (4) from equation (3) by means of a general inversion principle. However it was more expedient to prove (4) directly.

(2) Equations (3) and (4) show that the general Eisenstein series are modular forms of weight k for Γ_q and are linearly independent, as claimed.

(3) It now follows from our remarks at the beginning of §12, that every modular form can be written as a sum of general Eisenstein series and of Poincaré series φ_ν, $\nu \geq 1$.

§14. The Fourier Coefficients of Eisenstein Series

From now on, an "Eisenstein series" will be what was called in §13, a "general Eisenstein series".

As a first step towards computing the Fourier coefficients of the Eisenstein series, we prove

Theorem 1:

$$\sum_{n=-\infty}^{\infty} \frac{1}{(z+n)^r} = \frac{(-2\pi i)^r}{(r-1)!} \sum_{\nu=1}^{\infty} \nu^{r-1} e^{2\pi i \nu z}$$

if $r > 1$ and $\operatorname{Im} z > 0$.

Proof: We start with the well known partial fraction expansion for the cotangent function:

$$\pi \cot \pi z = \lim_{m \to \infty} \sum_{-m}^{m} \frac{1}{(z+n)} .$$

Differentiating term by term, and noticing that the differentiated series converges absolutely uniformly on compact sets, we obtain

$$\frac{d}{dz} (\pi \cot \pi z) = -\lim_{m \to \infty} \sum_{-m}^{m} \frac{1}{(z+n)^2} = -\sum_{-\infty}^{\infty} \frac{1}{(z+n)^2} ,$$

and similarly

$$\frac{d^{r-1}}{dz^{r-1}}(\pi \cot \pi z) = (-1)^{r-1} (r-1)! \sum_{n=-\infty}^{\infty} \frac{1}{(z+n)^r} .$$

On the other hand, we have

$$\pi \cot \pi z = -i\pi \frac{1 + e^{2\pi i z}}{1 - e^{2\pi i z}}$$

$$= -i\pi(1 + e^{2\pi i z})\left(\sum_{\nu=0}^{\infty} e^{2\pi i \nu z}\right), \quad \text{for} \quad \operatorname{Im} z > 0 ,$$

$$= (-\pi i)\left(1 + 2 \sum_{\nu=1}^{\infty} e^{2\pi i \nu z}\right) .$$

The series above converges absolutely uniformly on compact subsets of H and hence can be differentiated term by term to secure that

$$\frac{d^{r-1}}{dz^{r-1}}(\pi \cot \pi z) = -(2\pi i)^r \sum_{\nu=1}^{\infty} \nu^{r-1} e^{2\pi i \nu z} .$$

The desired result now follows upon comparing these two expansions.

<u>Theorem 2</u>: Let

$$A = \begin{cases} 0 & \text{if} \quad \gamma \not\equiv 0 \pmod q , \\ \displaystyle\sum_{\substack{n \equiv \delta \pmod q \\ n \neq 0}} \frac{1}{n^{2k}} & \text{if} \quad \gamma \equiv 0 \pmod q , \end{cases}$$

and

$$\sigma_r(\lambda; \gamma, \delta) = \sum_{\substack{d \mid \lambda \\ \frac{\lambda}{d} \equiv \gamma \pmod q}} d^r \cos \left(\frac{2\pi i d\delta}{q}\right) ;$$

then

$$G_k(z;\gamma,\delta;q) = A + \frac{(-1)^k(2\pi)^k 2}{q^{2k}(2k-1)!} \sum_{\lambda=1}^{\infty} \sigma_{2k-1}(\lambda;\gamma,\delta)e^{2\pi i\lambda z/q} .$$

We shall not prove the above theorem in full generality, but only for the full modular group Γ, since we shall use it only in that case. The proof is the same in general, though the notation becomes more forbidding. In the case $q = 1$, there is only one Eisenstein series, corresponding to $\gamma = 1$, $\delta = 0$, which we write as $G_k(z)$. The restricted Eisenstein series will be denoted by $G_k^*(z)$.

Theorem 2':

$$G_k(z) = 2\zeta(2k) + \frac{2(-1)^k(2\pi)^k}{(2k-1)!} \sum_{\lambda=1}^{\infty} \sigma_{2k-1}(\lambda)e^{2\pi i\lambda z} ,$$

where

$$\zeta(x) = \sum_{n=1}^{\infty} \frac{1}{n^x}$$

is the Riemann zeta function and

$$\sigma_r(\lambda) = \sigma_r(\lambda;1,0) = \sum_{d|\lambda} d^r .$$

Proof:

$$G_k(z) = \sum_{c=-\infty}^{\infty} {\sum_{d=-\infty}^{\infty}}' \frac{1}{(cz+d)^{2k}} \qquad \text{(we omit } c = d = 0)$$

$$= \sum_{\substack{d=-\infty \\ d\neq 0}}^{\infty} \frac{1}{d^{2k}} + \sum_{c=1}^{\infty} \sum_{d=-\infty}^{\infty} \frac{1}{(cz+d)^{2k}} + \sum_{c=-\infty}^{-1} \sum_{d=-\infty}^{\infty} \frac{1}{(cz+d)^{2k}} .$$

In the last sum, let $c' = -c$, $d' = -d$, to show that it is equal to the second sum; hence

$$G_k(z) = 2\zeta(2k) + 2 \sum_{c=1}^{\infty} \sum_{d=-\infty}^{\infty} \frac{1}{(cz+d)^{2k}} .$$

Since $c > 0$, $\mathrm{Im}(cz) > 0$, and Theorem 1 is applicable; then

$$G_k(z) = 2\zeta(2k) + \frac{2(-1)^k(2\pi)^{2k}}{(2k-1)!} \sum_{c=1}^{\infty} \sum_{\nu=1}^{\infty} \nu^{2k-1} e^{2\pi i\nu cz} .$$

Collecting terms for which $\nu c = \lambda$ gives

$$G_k(z) = 2\zeta(2k) + \frac{2(-1)^k(2\pi)^{2k}}{(2k-1)!} \sum_{\lambda=1}^{\infty} e^{2\pi i\lambda z} \left(\sum_{\nu\mid\lambda} \nu^{2k-1} \right) .$$

§15. The Modular Forms for the Modular Group

We recall the classical result that

(5) $$\zeta(2n) = \frac{B_n 2^{2n-1} \pi^{2n}}{(2n)!} ,$$

where the B_n are the Bernouilli numbers, defined for instance by:

$$z \cot z = 1 - \sum_{n=1}^{\infty} \frac{B_n 2^{2n} z^{2n}}{(2n)!} .$$

For future reference, we draw up a short table:

n	B_n	$\zeta(2n)$	n	B_n	$\zeta(2n)$
1	$\frac{1}{6}$	$\frac{\pi^2}{6}$	4	$\frac{1}{30}$	$\frac{\pi^8}{2\cdot 3^3\cdot 5^2\cdot 7}$
2	$\frac{1}{30}$	$\frac{\pi^4}{90}$	5	$\frac{5}{66}$	$\frac{\pi^{10}}{3^5\cdot 5\cdot 7\cdot 11}$
3	$\frac{1}{42}$	$\frac{\pi^6}{3^3\cdot 5\cdot 7}$	6	$\frac{691}{2730}$	$\frac{\pi^{12}}{3^6\cdot 5^3\cdot 7^2\cdot 11\cdot 13}$

By Theorem 2' and (5),

$$E_k(z) = \frac{1}{2\zeta(2k)} G_k(z) = 1 + \frac{(-1)^k 4k}{B_k} \sum_{n=1}^{\infty} \sigma_{2k-1}(n) e^{2\pi i n z}$$

is a modular form of weight k for Γ, called the normalized Eisenstein series. For instance:

$$E_2(z) = 1 + 240 \sum_{n=1}^{\infty} \sigma_3(n) e^{2\pi i n z} \quad ,$$

$$E_3(z) = 1 - 504 \sum_{n=1}^{\infty} \sigma_5(n) e^{2\pi i n z} \quad ,$$

$$E_4(z) = 1 + 480 \sum_{n=1}^{\infty} \sigma_7(n) e^{2\pi i n z} \quad ,$$

$$E_5(z) = 1 - 264 \sum_{n=1}^{\infty} \sigma_9(n) e^{2\pi i n z} \quad ,$$

$$E_6(z) = 1 + \frac{54600}{691} \sum_{n=1}^{\infty} \sigma_{11}(n) e^{2\pi i n z} \quad .$$

In §8 we saw that the dimension of the space of modular forms of weight k for Γ is:

$$\delta_k = \begin{cases} \left[\frac{k}{6}\right] & \text{if } k \equiv 1 \pmod 6 , \\ \left[\frac{k}{6}\right] + 1 & \text{if } k \not\equiv 1 \pmod 6 . \end{cases}$$

In particular, for $k = 2,3,4,5,7$, all modular forms of weight k are multiples of E_2, E_3, E_4, E_5 and E_7 respectively, since in those cases $\delta_k = 1$. In other words, for those values of k, there are no cusp forms. For $k = 6$, there are two modular forms, one of which must be a cusp form. Now $(E_2(z))^3$ and $(E_3(z))^2$ are both forms of weight 6, and both are one at $i\infty$; hence

(7) $$\Delta(z) = \frac{64\pi^{12}}{27} (E_2(z)^3 - E_3(z)^2)$$

is a cusp form of weight 6. We see that $\Delta(z)$ does not vanish identically by computing the first few terms in its Fourier expansion:

$$\Delta(z) = (2\pi)^{12} (e^{2\pi iz} - 24e^{4\pi iz} + 252e^{6\pi iz} + \ldots) ,$$

or

$$\Delta(z) = (2\pi)^{12} \left(\sum_{n=1}^{\infty} \tau(n) e^{2\pi inz} \right) ,$$

where the coefficients $\tau(n)$ are called Ramanujan's numbers.

Theorem 3: $\Delta(z)$ has the product expansion:

$$\Delta(z) = (2\pi)^{12} e^{2\pi iz} \left\{ \prod_{n=1}^{\infty} (1-e^{2\pi inz}) \right\}^{24} .$$

For the proof of this result, the reader is referred to Hurwitz' thesis [1], or a very short proof by C. L. Siegel [2].

Theorem 4: Let us write

$$\sigma_{2r-1}(0) = \frac{1}{2} \zeta(1-2r) = \frac{\Gamma(2r)\, \zeta(2r)(-1)^r}{(2\pi)^{2r}} ;$$

then, if $k, \ell > 1$,

(8) $$\sum_{m=0}^{n} \sigma_{2k-1}(m)\sigma_{2\ell-1}(n-m) = \frac{\Gamma(2k)\,\Gamma(2\ell)}{(2(k+\ell))} \frac{\zeta(2k)\,\zeta(2\ell)}{\zeta(2(k+\ell))} \sigma_{2(k+\ell)-1}(n) + a_n$$

where a_n is the nth-Fourier coefficient of some cusp form of weight $k + \ell$. In particular, $a_n = o(n^{k+\ell})$ always. (Note that the main term is $o(n^{2(k+\ell)-1})$.

Proof: Since $E_r(z)$ is a modular form of weight r for Γ and $E_r(i\infty) = 1$, we see that

$$E_k(z)\, E_\ell(z) - E_{k+\ell}(z) = \varphi_{k+\ell}(z)$$

is a cusp form of weight $k + \ell$. Comparing the coefficients leads to the result.

Corollary 1: If $r = 2,3,4,5,7$, there are no cusp forms of weight r. In particular:

(1) $(E_2(z))^2 = E_4(z)$, hence

$$\sigma_7(n) = \sigma_3(n) + 120 \sum_{m=1}^{n-1} \sigma_3(m)\, \sigma_3(n-m)$$

by the above theorem;

(2) $E_2(z)\, E_3(z) = E_5(z)$, hence

$$11\sigma_9(n) = 21\sigma_5(n) - 10\sigma_3(n) + 5040 \sum_{m=1}^{n=1} \sigma_3(n)\sigma_5(n-m) \quad ;$$

(3) $E_3(z)\, E_4(z) = E_2(z)\, E_5(z) = E_7(z)$, which provide some more arithmetical identities.

Corollary 2: In thoerem 3, let $k = \ell = 3$; then $a_n = c\tau(n)$ for some constant c, since there is a unique cusp form of weight 6, namely $\Delta(z)$. Evaluating c, we obtain

$$\tau(n) = \frac{65}{756}\, \sigma_{11}(n) + \frac{691}{756}\, \sigma_5(n) - \frac{691}{3} \sum_{m=1}^{n-1} \sigma_5(m)\, \sigma_5(n-m) \quad .$$

Remark: Ramanujan proved formula (8) without explicit reference to modular forms, by means of generating series similar to the normalized Eisenstein series. However, his error term is not as good as ours, his being:

$$a_n = O\left(n^{\frac{4}{3}(k+\ell)-\frac{2}{3}}\right) \quad .$$

Theorem 5: All modular forms for Γ are polynomials in $E_2(z)$ and $E_3(z)$.

Proof: Clearly the theorem is true for $k = 2,3,4,5,6,7$. We have seen that $\delta_{k+6}(\Gamma) = \delta_k(\Gamma) + 1$, where $\delta_k(\Gamma)$ is the space of modular forms of weight k for Γ. Assume the theorem true for k; we show that it is true for $k + 6$. Let a basis for the modular forms of weight k be $\{f_1, \ldots, f_{\delta_k}\}$, and let $E_2^{\alpha} E_3^{\beta} = f_{(\delta_k)+1}$ be a modular form of weight $k + 6$. Then

$$\{\Delta f_1, \Delta f_2, \ldots, \Delta f_{\delta_k}, f_{(\delta_k)+1}\}$$

forms a basis for the space of modular forms of weight $k + 6$; since each member of the basis is a polynomial in $E_2(z)$ and $E_3(z)$, the induction is completed.

BIBLIOGRAPHY

[1]. Hurwitz, A., Math. Ann., vol. 18 (1881), pp. 528-592.

[2]. Siegel, C. L., Matematika, vol. 1 (1954) pp. 4.

CHAPTER V:

MODULAR CORRESPONDENCES

§16. The Hecke Operators

In this chapter we prove some arithmetic theorems about the Fourier coefficients of cusp forms. For simplicity we restrict ourselves to the full modular group Γ. The results are obtained by studying an algebra of operators introduced by Hecke [1].

Let T be any real linear fractional transformation. Let $\mathscr{A}$ be the vector space of holomorphic functions on the upper half plane. We define an operator $T\colon \mathscr{A} \longrightarrow \mathscr{A}$ by

$$(f|_k T)(z) = J_T(z)^k \, f(Tz) \quad .$$

Clearly the operator T is linear and $f|_k(T_1T_2) = (f|_kT_1)|_kT_2$. If f is a modular form of weight k for a group G, then $f|_kT = f$ for all T in G.

Suppose G is a subgroup of Γ, and $H \supset G$ is a set of linear fractional transformations (not necessarily a group) with the following properties:

(1) $HG = GH = H$;

(2) $[H:G] < \infty$, i.e., $H = \cup_1^m GM_j$ for some elements M_j in H .

Let $\Delta(\mathscr{S}')$ be the group of divisors on $\mathscr{S}$, the Riemann surface of (see §7). We introduce for each such set H an endomorphism T of $\Delta(\mathscr{S})$ defined on the generators of $\Delta(\mathscr{S})$ by

$$T(\{z\}) = \sum_{i=1}^{m} \{M_i\, z\} \quad ,$$

where z is a point of the upper half-plane (or a parabolic vertex),

$\{z\}$ denotes its equivalence class under G, and $\{M_1, M_2, \dots, M_m\}$ is a complete set of right coset representatives of $H \bmod G$. It follows from our assumptions that the operation is well-defined and is independent of the choice of coset representatives. From the endomorphism T defined above we also construct an operator T on the space of modular forms of weight k by defining

$$f \circ T = \sum_{i=1}^{m} f|_k M_i \quad .$$

<u>Remarks</u>: (1) $f \circ T$ is independent of the choice of coset representatives $\{M_i\}$ of $H \bmod G$. If $\{M_i'\}$ is another such set, then $M_i' = S_i M_i$ with S_i in G; hence

$$\sum_{i=1}^{m} f|_k M_i' = \sum_{i=1}^{m} f|_k (S_i M_i) = \sum_{i=1}^{m} (f|_k S_i)|_k M_i = \sum_{i=1}^{m} f|_k M_i \quad ,$$

since f is a modular form for G. (2) $f \circ T$ is an unrestricted modular form of weight k for G. Let S be in G; then

$$\begin{aligned}(f \circ T)|_k S &= \Big(\sum f|_k M_i\Big)\Big|_k S \\ &= \sum f|_k M_i S \\ &= \sum f|_k M_i = f \circ T \quad ,\end{aligned}$$

since $\{M_i S\}$ is a complete set of coset representatives for $H \bmod G$ whenever $\{M_i\}$ is.

From now on, in this chapter, we restrict ourselves to the case that G is the inhomogeneous modular group Γ, and consider the endomorphisms T_n arising from the sets

$$H_n = \left\{ \begin{pmatrix} a & b \\ c & d \end{pmatrix} \middle| \begin{matrix} ad - bc = n \\ a,b,c,d \text{ integers} \end{matrix} \right\} \quad .$$

Property (1) is fulfilled, as is seen by taking determinants; Property (2) follows from

<u>Lemma 1</u>: The set M of linear fractional transformations

represented by the set of matrices

$$M' = \left\{ \begin{pmatrix} a & b \\ 0 & d \end{pmatrix} \middle| ad = n \ , \ d > 0 \ , \ b \bmod d \right\}$$

form a complete set of right coset representatives of H modulo Γ.

<u>Proof</u>: Let $\pm\begin{pmatrix} \alpha & \beta \\ \gamma & \delta \end{pmatrix}$ be in Γ and $\pm\begin{pmatrix} A & B \\ C & D \end{pmatrix}$ be in H_n; then

$$\pm\begin{pmatrix} \alpha & \beta \\ \gamma & \delta \end{pmatrix}\begin{pmatrix} A & B \\ C & D \end{pmatrix} = \pm\begin{pmatrix} * & * \\ A\gamma+C\delta & * \end{pmatrix} ,$$

from which it is clear that we can assume that the right coset representatives of H_n mod Γ are triangular matrices, hence that

$$M' \subseteq \left\{ \begin{pmatrix} A & C \\ 0 & D \end{pmatrix} \middle| \ A,C,D \text{ integers}, \ D > 0, \ AD = n \right\} .$$

Note that $\begin{pmatrix} A & B \\ 0 & D \end{pmatrix}$ and $\begin{pmatrix} A' & B' \\ 0 & D' \end{pmatrix}$ are in the same right coset mod Γ if and only if $A = A'$, $D = D'$ and $B \equiv B' \bmod D$.

Let $\mathcal{H}$ be the algebra generated by the endomorphisms T_n, $n = 1,2,3,\ldots$.

<u>Theorem 1</u>: $\mathcal{H}$ is a commutative algebra, generated by the operators T_p for all prime numbers p. In fact:

(1)
$$T_n \, T_m = \sum_{d|(n,m)} d \, T_{\frac{nm}{d^2}} .$$

<u>Proof:</u> (a) If $(m,n) = 1$, then $T_m \, T_n = T_{mn}$. For by Lemma 1,

$$T_m(z) = \sum_{\substack{ad=m \\ d>0 \\ b \bmod d}} \left\{\frac{az+b}{d}\right\} , \quad T_n(z) = \sum_{\substack{a'd'=n \\ d'>0 \\ b' \bmod d'}} \left\{\frac{a'z+b'}{d'}\right\} ,$$

hence

$$T_n \, T_m = \sum \left\{ \frac{aa'z + (ab' + bd')}{dd'} \right\} .$$

Since m and n are coprime, it is sufficient to show that when b (resp. b') runs through a residue system mod d (resp. d'), then $ab' + bd'$ (for fixed a and d') runs through a residue system mod dd'. But there are dd' such numbers; and since they are incongruent mod dd', they must form a residue system for dd'. Thus $T_n T_m = T_{nm}$.

(b) Let p be a prime and $r > 0$. Then

$$T_p \cdot T_{p^r} = T_{p^{r+1}} + pT_{p^{r-1}} .$$

For from Lemma 1,

$$T_{p^r}(z) = \sum_{\substack{0 \le i \le r \\ b_i \bmod p^i}} \left\{\frac{p^{r-i}z+b_i}{p^i}\right\} ,$$

$$T_p(z) = \{pz\} + \sum_{b \bmod p} \left\{\frac{z+b}{p}\right\} ,$$

so that

$$T_p T_{p^r}(z) = \sum_{\substack{0 \le i \le r \\ b_i \bmod p^i}} \left\{ \frac{p^{r+1-i}r+b_i p}{p^i} \right\} + \sum_{\substack{0 \le i \le r \\ b_i \bmod p^i \\ b \bmod p}} \left\{ \frac{p^{r-i}z+b_i+bp^i}{p^{i+1}} \right\} .$$

In the second sum, $b_i + bp^i$ runs through a complete residue system mod p^{i+1}, so that

$$T_p T_{p^r}(z) = \sum_{\substack{0 \le i \le r \\ b_i \bmod p^i}} \left\{ \frac{p^{r+1-i}z+b_i p}{p^i} \right\} + T_{p^{r+1}}(z) - \{p^{r+1}z\}$$

$$= \sum_{\substack{1 \le i \le r \\ b_i \bmod p^i}} \left\{ \frac{p^{r+1-i}z+b_i p}{p^i} \right\} + T_{p^{r+1}}(z) .$$

In the sum let $b_i = b_{i-1} + p^{i-1}b$, where b_{i-1} runs through a residue system mod p^{i-1} and b through a system mod p. As points of $\mathcal{S}$, $\{w\}$ and $\{w+b\}$ are the same, so that each term $\left\{ \dfrac{p^{r+1-i}z+b_{i-1}p}{p} \right\}$ occurs exactly p times; thus

$$T_p T_{p^r}(z) = T_{p^{r+1}}(z) + p \sum_{\substack{0 \le i \le r-1 \\ b_i \bmod p^i}} \left\{ \frac{p^{r-1-i}z+b_i}{p^i} \right\} = T_{p^{r+1}}(z) + p\, T_{p^{r-1}}(z)$$

(c) An easy induction shows that if $r, s > 0$

$$T_{p^r} T_{p^s} = \sum_{t=0}^{\min(r,s)} p^t \, T_{p^{(r+s-2t)}} ,$$

from which (a) shows that

$$T_m T_n = \sum_{d \mid (m,n)} d \, T_{\frac{mn}{d^2}} .$$

Let $\mathcal{M}_k$ be the space of modular forms of weight k for G.

Corollary: Let $\mathcal{T}$ be the algebra of operators on $\mathcal{M}_k$ induced by the endomorphisms of $\mathcal{H}$. The natural map of $\mathcal{H}$ to $\mathcal{T}$ is clearly a homomorphism from the algebra $\mathcal{H}$ onto the algebra $\mathcal{T}$, so that $\mathcal{T}$ is a commutative algebra. The generators T_n satisfy the relations

$$T_m T_n = \sum_{d \mid (m,n)} d \, T_{\frac{mn}{d^2}} .$$

Proof: We have seen that T_n takes modular forms of weight k into unrestricted modular forms of the same weight. It is sufficient to check the behaviour at $i\infty$ of the image. We have

$$(f \circ T_n)(z) = \sum_{\substack{ad=n \\ d>0 \\ b \bmod d}} \left(f \Big|_k \begin{pmatrix} a & b \\ 0 & d \end{pmatrix} \right)(z)$$

$$= \sum_{\substack{ad=n \\ d>0 \\ b \bmod d}} \left(\frac{a}{d} \right)^k f\left(\frac{az+b}{d} \right) ,$$

so that

(2) $$(f \circ T_n)(z) = n^k \sum_{\substack{ad=n \\ d>0 \\ b \bmod d}} d^{-2k} f\left(\frac{az+b}{d} \right) .$$

This is a finite sum of functions holomorphic at $i\infty$, and so is also holomorphic there. In particular, if f is a cusp form so is $f \circ T_n$.

<u>Theorem 2</u>: Let $k > 1$ and

$$\varphi_\nu(z) = \sum_{T \text{ in } \mathcal{R}} e^{2\pi i \nu T z} J_T(z)^k \qquad \nu = 0, 1, \ldots,$$

be the Poincaré series of weight k for Γ. Then

$$\varphi_\nu \circ T_n = n^k \sum_{d \mid (n,\nu)} d^{1-2k} \varphi_{\frac{\nu n}{d^2}} \quad . \tag{3}$$

<u>Proof</u>: (a) We recall that $\mathcal{R}$ is a set of right coset representatives of $\Gamma \bmod \Gamma_0$ (where Γ_0 is the subgroup of all translations). Let $\mathcal{Q}$ be a set of right coset representatives of $H_n \bmod \Gamma$. Then

$$(\varphi_\nu \circ T_n)(z) = \sum_{M \text{ in } \mathcal{Q}} J_M(z)^k \varphi_\nu(M z)$$

$$= \sum_{M \text{ in } \mathcal{Q}} \sum_{T \text{ in } \mathcal{R}} e^{2\pi i \nu\, TMz} \left(J_T(Mz) J_M(z)\right)^k$$

$$= \sum_{S \text{ in } \mathcal{P}} e^{2\pi i \nu\, Sz} J_S(z)^k \quad ,$$

where $\mathcal{P} = \mathcal{R}\mathcal{Q}$ is a set of right coset representatives for $H_n \bmod \Gamma_0$. We observe that the sum is independent of our choice $\mathcal{P}$ of coset representatives.

(b) We now check that $\mathcal{Q}\mathcal{R}$ is a set of right coset representatives of $H_n \bmod \Gamma_0$. Let $\begin{pmatrix} A & B \\ C & D \end{pmatrix}$, (with $AD - BC = n$), be in H_n. We wish to write

$$\underset{\varepsilon\, \Gamma_0}{\begin{pmatrix} 1 & q \\ 0 & 1 \end{pmatrix}} \underset{\varepsilon\, \mathcal{Q}}{\begin{pmatrix} a & b \\ 0 & d \end{pmatrix}} \underset{\varepsilon\, \mathcal{R}}{\begin{pmatrix} \alpha & \beta \\ \gamma & \delta \end{pmatrix}} = \begin{pmatrix} A & B \\ C & D \end{pmatrix} \tag{4}$$

First, we can find $\gamma, \delta,$ such that $(\gamma, \delta) = 1$ and

$$\begin{pmatrix} A & B \\ C & D \end{pmatrix} \begin{pmatrix} \delta & -\beta \\ -\gamma & \alpha \end{pmatrix} = \begin{pmatrix} * & * \\ C\delta - \gamma D & * \end{pmatrix}$$

is a triangular matrix, i.e., such that $C\delta = \gamma D$. Let $\begin{pmatrix} \alpha & \beta \\ \gamma & \delta \end{pmatrix}$ in $\mathcal{R}$ be the unique matrix with second row $\{\gamma, \delta\}$. Multiplying both sides of (4) by

$\begin{pmatrix}\alpha & \beta \\ \gamma & \delta\end{pmatrix}^{-1}$ shows that we can assume that $C = 0$. But then we need only take $a = A$, $d = D$ and q such that $B - qD = b$ is in our residue system mod D.

(c) We now return to part (a).

$$\varphi_\nu \circ T_n = \sum_{S \text{ in } \mathcal{P}} e^{2\pi i\nu\, Sz}\, J_S(z)^k$$

$$= \sum_{S = MT \in \mathcal{Q}\mathcal{R}} e^{2\pi i\nu\, MTz}\, J_M(Tz)^k\, J_T(z)^k$$

$$= \sum_{\substack{ad=n \\ d>0}} \left(\frac{a}{d}\right)^k \sum_{T \text{ in } \mathcal{R}} e^{2\pi i\nu\frac{a}{d}}\, J_T(z)^k \sum_{b \bmod d} e^{2\pi i\nu\frac{b}{d}} \quad ;$$

and since the innermost sum is

$$\sum_{b \bmod d} e^{2\pi i\nu\frac{b}{d}} = \begin{cases} d & \text{if } d\,|\,\nu \\ 0 & \text{otherwise} \end{cases} ,$$

it follows that

$$\varphi_\nu \circ T_n = n^k \sum_{d\,|\,(n,\nu)} d^{1-2k} \sum_{T \text{ in } \mathcal{R}} e^{2\pi i(\frac{\nu n}{d^2})Tz}\, J_T(z)^k \quad .$$

<u>Corollary 1</u>: $\varphi_0 \circ T_n = n^{1-k}\, \sigma_{2k-1}(n)\, \varphi_0$, so that the Eisenstein series is an eigenfunction for $\mathcal{T}$.

<u>Corollary 2</u>: $\varphi_1 \circ T_n = n^k\, \varphi_n$.

<u>Corollary 3</u>: $n^{-k}\, \varphi_\nu \circ T_n = \nu^{-k}\, \varphi_n \circ T_\nu$.

<u>Proof</u>:

$$n^{-k}\, \varphi_\nu \circ T_n = \sum_{d\,|\,(n,\nu)} d^{1-2k}\, \varphi_{\frac{n\nu}{d^2}}$$

is symmetric in n and ν .

§ 17. <u>Hecke Operators and Fourier Coefficients</u>.

Let

$$\varphi_\nu(z) = \sum_{\lambda=1} c_\lambda(\nu) e^{2\pi i\lambda z}, \quad \nu \geq 1,$$

be the νth Poincaré series and

$$(1) \qquad n^{-k}\, \varphi_\nu(z)\circ T_n = \sum_{\lambda=1}^{\infty} c_\lambda(\nu,n) e^{2\pi i\lambda z} \quad ,$$

where c_λ are the Fourier coefficients. The results of the last section lead to relations between these Fourier coefficients.

Theorem 2':

$$c_\lambda(\nu,n) = \sum_{d\,|\,(\nu,n)} d^{1-2k}\, c_\lambda\left(\frac{n\nu}{d^2}\right) \quad .$$

Corollary 3': $c_\lambda(\nu,n) = c_\lambda(n,\nu)$. [A]

Theorem 3: Let

$$f(z) = \sum_{\lambda=1}^{\infty} s_\lambda e^{2\pi i\lambda z}$$

be an arbitrary modular form of weight $k > 1$, and

$$f(z)\circ T_n = \sum_{\lambda=0}^{\infty} s_\lambda(n) e^{2\pi i\lambda z} \quad .$$

Then

$$s_\lambda(n) = \begin{cases} n^{1-k} \displaystyle\sum_{d\,|\,(n,\lambda)} d^{2k-1}\, s_{\frac{\lambda n}{d^2}} & \text{if } \lambda \geq 1 \;, \\ s_0\, \sigma_{2k-1}(n)\, n^{1-k} & \text{if } \lambda = 0 \;. \end{cases}$$

Proof: We have seen that

$$n^{-k} f(z)\circ T_n = \sum_{\substack{ad=n \\ d>0 \\ b \bmod d}} d^{-2k} f\left(\frac{az+b}{d}\right)$$

$$(2) \qquad = \sum_{\substack{ad=n \\ d>0 \\ b \bmod d}} d^{-2k} \left(\sum_{\lambda=1}^{\infty} s_\lambda e^{2\pi i\frac{\lambda az}{d}} e^{2\pi i\frac{\lambda b}{d}} + s_0 \right)$$

$$= s_0 n^{1-2k} \sigma_{2k-1}(n) + \sum_{\substack{ad=n \\ d>0}}^{\infty} \sum_{\lambda=1}^{\infty} d^{-2k} s_\lambda e^{2\pi i\lambda az/d} \sum_{b \bmod d} e^{2\pi ib\lambda/d}$$

The last sum, over b, is d or 0 according as d divides λ or not; so letting $\lambda = \mu d$,

$$n^{-k}f(z)\circ T_n = s_0 n^{1-2k}\sigma_{2k-1}(n) + n^{1-2k}\sum_{a|n}\sum_{\mu=1}^{\infty} a^{2k-1}\, s_{\frac{\mu n}{a}}\, e^{2\pi i\mu a z} .$$

Collecting terms with $\lambda = \mu a$

$$n^{-k}f(z)\circ T_n = s_0 n^{1-2k}\sigma_{2k-1}(n) + n^{1-2k}\sum_{\lambda=1}^{\infty} e^{2\pi i\lambda z}\Big(\sum_{a|(n,\lambda)} a^{2k-1}\, s_{\frac{\mu n}{a^2}}\Big) .$$

Comparing coefficients on both sides proves the theorem.

Corollary 4: In the above notation we have

$$c_\mu(\nu,n) = n^{1-2k}\sum_{d|(n,\mu)} d^{2k-1}\, c_{\frac{\mu n}{d^2}}(\nu) \quad ;$$

and in particular we have the symmetry law

$$\mu^{1-2k}c_\mu(\nu,n) = n^{1-2k}\, c_n(\nu,\mu) \quad . \qquad \boxed{B}$$

We have now gathered sufficient information about the Hecke operators and their action on the Poincaré series to prove that they are Hermitian operators on the space of cusp forms under the Petersson inner product.

Theorem 4: $(\varphi_\nu\circ T_n,\ \varphi_\mu) = (\varphi_\mu\circ T_n,\ \varphi_\nu)$. $\boxed{C}$

Proof: By Theorem 5 of §11, if f is any cusp form and a_μ is its μth coefficient then

$$(f,\varphi_\mu) = \rho_k\, \mu^{1-2k}\, a_\mu \quad ,$$

where ρ_k is a real constant. Recalling that

$$\varphi_\nu\circ T_n = n^k\sum_{\lambda=1}^{\infty} c_\lambda(\nu,n)e^{2\pi i\lambda z} \quad ,$$

we have

$$(\varphi_\nu\circ T_n,\ \varphi_\mu) = \rho_k\, \mu^{1-2k}\, n^k\, c_\mu(\nu,n)$$

$$= \rho_k\, n^{1-2k}\, n^k\, c_n(\nu,\mu) \qquad \text{by B}$$

$$= \rho_k \, n^{1-2k} \, n^k \, c_n(\mu,\nu) \qquad \text{by A}$$

$$= \rho_k \, \nu^{1-2k} \, n^k \, c_\nu(\mu,n) \qquad \text{by B}$$

$$= (\varphi_\mu \circ T_n, \varphi_\nu) \; .$$

Theorem 5: $\mathcal{T}$ is a commutative algebra of Hermitian operators on the Hilbert space of cusp forms.

Proof: It suffices to prove that T_n is Hermitian for all n. Substituting $n = 1$ in C gives, since $T_1 = I$,

$$(\varphi_\nu \circ T_1, \varphi_\mu) = (\varphi_\nu, \varphi_\mu) = (\varphi_\mu, \varphi_\nu) = \overline{(\varphi_\nu, \varphi_\mu)} \; .$$

Since $(\varphi_\nu, \varphi_\mu) = \rho_k \, \mu^{1-2k} \, c_\nu(\mu)$, this shows that the coefficients of the Poincaré series, $c_\nu(\mu)$, are all real. Now, by Theorem 2'

$$(\varphi_\nu \circ T_n, \varphi_\mu) = \varphi_k \, \mu^{1-2k} \, n^k \sum_{d \mid (n,\nu)} d^{1-2k} \, c_\mu\left(\frac{n\nu}{d^2}\right) \; ,$$

which shows that $(\varphi_\nu \circ T_n, \varphi_\mu)$ is real for all n, ν, μ. Thus

$$\begin{aligned} (\varphi_\nu \circ T_n, \varphi_\mu) &= (\varphi_\mu \circ T_n, \varphi_\nu) \qquad \text{by C} \\ &= \overline{(\varphi_\nu, \varphi_\mu \circ T_n)} \\ &= (\varphi_\nu, \varphi_\mu \circ T_n) \; . \end{aligned}$$

The Poincaré series generate the cusp forms. Let $f = \Sigma \, a_\nu \varphi_\nu$, $g = \Sigma \, b_\mu \varphi_\mu$ be two cusp forms; then

$$(f \circ T_n, g) = \sum_{\nu,\mu} a_\nu \, \overline{b}_\mu (\varphi_\nu \circ T_n, \varphi_\mu)$$

$$= \sum_{\nu,\mu} a_\nu \, \overline{b}_\mu (\varphi_\nu, \varphi_\mu \circ T_n)$$

$$= (f, g \circ T_n) \; ,$$

proving that T_n is Hermitian.

§18. Arithmetic Properties of the Fourier Coefficients.

In this section we apply the results obtained in the last two paragraphs to the study of the Fourier coefficients of cusp forms.

Let $f = \{f_1, \ldots, f_r\}$ be an orthogonal basis for the Hilbert space of cusp forms. Then $f_i \circ T_n = \sum_j f_j \lambda_{ij}(n)$, or, in matrix notation

$$\underline{f} \circ T_n = \underline{f}\, \Lambda(n) \quad \text{with} \quad \Lambda(n) = (\lambda_{ij}(n)) \quad .$$

The map $T_n \longrightarrow \Lambda(n)$ generates an algebra homomorphism. Let $\mathcal{L}$ be the algebra of matrices generated by the $\Lambda(n)$. It is clear that $\Lambda(n)$ is Hermitian; for

$$(f_i \circ T_n, f_j) = \sum_k \lambda_{ik}(n)(f_k, f_j) = \lambda_{ij}(n)$$

$$= (f_i, f_j \circ T_n) = \sum_k \overline{\lambda_{jk}(n)}(f_i, f_k) = \overline{\lambda_{ji}(n)}$$

We can now prove the important:

<u>Theorem</u> (Hecke): We can select an orthogonal basis for the cusp forms of weight k for Γ for which $\mathcal{L}$ is an algebra of diagonal matrices. The functions of the basis have "multiplicative" Fourier coefficients.

<u>Proof</u>: We have seen that $\mathcal{L}$ is a commutative algebra of Hermitian matrices. It can therefore be diagonalized by an appropriate choice of orthogonal basis. Let $\{g_1, \ldots, g_r\}$ be such a basis; then clearly each g_i is an eigenfunction for all T_n. We can assume that the first Fourier coefficient of $g_i (i=1, \ldots r)$ is 1. It remains to show that the Fourier coefficients of g_i are multiplicative. Let

$$g_i(z) = \sum_{\lambda=1}^{\infty} a_i(\lambda) e^{2\pi i \lambda z}$$

and

$$g_i(z) \circ T_n = t_i(n)\, g_i(z) \quad ;$$

Theorem 3 shows that

$$t_i(n)\, a_i(\lambda) = n^{1-k} \sum_{d \mid (n,\lambda)} d^{2k-1}\, a_i\left(\frac{\lambda n}{d^2}\right) \quad .$$

If $\lambda = 1$, $a_i(\lambda) = 1$ and $t_i(n) = n^{1-k}\, a_i(n)$; thus

$$a_i(n)\, a_i(\lambda) = \sum_{d \mid (n,\lambda)} d^{2k-1}\, a_i\left(\frac{\lambda n}{d^2}\right) \quad ,$$

and in particular, if $(n,\lambda)=1$, $a_1(n)\,a_1(\lambda) = a_1(\lambda n)$.
This is the sense in which "multiplicative" is meant.

Remark: In the proof, we saw that if $t_1(n)$ is the eigenvalue of T_n to which g_1 belongs, then $t_1(n) = n^{1-k}\,a_1(n)$. Thus, to know the Fourier coefficients of cusp forms, it would be sufficient to know the eigenvalues of T_n, $n=1,2,\ldots$.

Corollary 5: The Ramanujan function $\tau(n)$ is multiplicative.

Proof: There is a unique cusp form of weight 6,

$$(2\pi)^{-12}\,\Delta(z) = \sum_{n=1}^{\infty} \tau(n)e^{2\pi inz} ;$$

Hecke's theorem shows that

$$\tau(n)\ \tau(m) = \sum_{d\mid(n,m)} \tau\!\left(\frac{nm}{d^2}\right)d^{11} .$$

To any Fourier series

$$f(z) = \sum_{n=1}^{\infty} a_n\ e^{2\pi inz}$$

we can formally associate the Dirichlet series

$$\varphi(s) = \sum_{n=1}^{\infty} \frac{a_n}{n^s} .$$

$\varphi(s)$ can be considered as the "Mellin transform" of $f(z)$, since

$$\varphi(s) = \frac{(2\pi)^s}{\Gamma(s)} \int_0^{\infty} f(iz)z^{s-1}dz .$$

If $f(z)$ is a cusp form of weight k, then $a_n = O(n^k)$ and the Dirichlet series $\varphi(s)$ converges to an analytic function of s for $Re(s) > k+1$. Let $\psi(s) = \Gamma(s)\,\varphi(s)(2\pi)^{-s}$. The relation $f(\frac{i}{z}) = (iz)^{2k}\,f(iz)$ satisfied by the cusp form $f(z)$ implies the functional equation $\psi(2k-s) = (-1)^k\,\psi(s)$.

Theorem 6: Let $f(z)$ be a cusp form, of weight k for Γ, which is an eigenfunction for $\mathcal{T}$. Then its "Mellin transform" $\varphi(s)$ has an

Euler product expansion

$$\varphi(s) = \prod_p \left(1-a_p p^{-s} + p^{2k-1-2s}\right)^{-1} .$$

Proof: We have

$$\varphi(s) = \sum_{n=1}^{\infty} \frac{a_n}{n^s} ,$$

where the coefficients a_n satisfy the condition

$$a_n\, a_m = \sum_{d\mid(n,m)} d^{2k-1}\, a_{\frac{nm}{d^2}}$$

Since, in particular, $a_n\, a_m = a_{nm}$ if $(n,m) = 1$, it follows that

$$\varphi(s) = \prod_p \left(\sum_{m=0}^{\infty} \frac{a_{p^m}}{p^{ms}} \right) .$$

Let us evaluate a typical factor

$$\varphi_p(s) = \sum_{m=0}^{\infty} \frac{a_{p^m}}{p^{ms}} .$$

$$\frac{a_p}{p^s}\, \varphi_p(s) = \sum_{m=0}^{\infty} \frac{a_{p^m}\, a_p}{p^{(m+1)s}} = \sum_{m=0}^{\infty} \frac{a_{p^{m+1}} + p^{2k-1}\, a_{p^{m-1}}}{p^{(m+1)s}} = \varphi_p(s) - 1 + p^{2k-1-2s}\varphi_p(s).$$

hus

$$\varphi_p(s) = (1-a_p\, p^{-s} + p^{2k-1-2s})^{-1} .$$

BIBLIOGRAPHY

[1]. Hecke, E , Math. Ann., vol. 114 (1937), pp. 1-28, 316-351.

CHAPTER VI:

QUADRATIC FORMS

We associate to every positive definite quadratic form an analytic function, its theta series. Under certain conditions the theta series is a modular form for a suitable subgroup of the modular group, and we can then apply the function-theoretic results we have obtained in order to get information about the representation numbers of the quadratic form.

§19. Introduction.

We shall use the standard matrix notation. Let $A = (a_{ij})$ be a real-valued $r \times r$ matrix, and ${}^tA$ denote its transpose: ${}^tA = (a_{ji})$. A is symmetric if ${}^tA = A$. By $X, Y, \ldots$ we mean the column vectors

$$X = \begin{pmatrix} x_1 \\ \vdots \\ x_r \end{pmatrix} \qquad Y = \begin{pmatrix} y_1 \\ \vdots \\ y_r \end{pmatrix}$$

If X, Y are two vectors, $X \cdot Y = \Sigma\, x_i y_i$ will be their scalar product.

To the matrix $A = (a_{ij})$ we associate the quadratic form

$$A[X] = X \cdot A\, X = \sum_{ij} a_{ij} x_i x_j \quad .$$

Since we are interested in the quadratic form rather than the matrix, we can assume without loss of generality that A is symmetric.

The quadratic form $A[X]$ is said to be positive definite if $A[X]$ is positive whenever $X \neq 0$. Diagonalizing A shows that $A[X]$ is positive definite if and only if there is a positive constant c such that

$$A[X] > c \sum_i x_i^2 \quad \text{for} \quad X = \begin{pmatrix} x_1 \\ \vdots \\ x_r \end{pmatrix} \neq 0 \ .$$

We are interested in quadratic forms with integer coefficients; thus we restrict A to be semi-integral, meaning that $2a_{ij}$, a_{ii} are integers.

Henceforth A will be a positive definite, symmetric, semi-integral matrix. To such a matrix A we associate the theta series

$$\theta_A(z) = \sum_N e^{\pi i A[N]z}, \quad (\text{Im } z > 0),$$

where N runs through all integral vectors $N = \begin{pmatrix} n_1 \\ \vdots \\ n_r \end{pmatrix}$.

Since A is positive definite, $\theta_A(z)$ is holomorphic on the upper half plane, as one sees immediately from the bound

$$|\theta_A(z)| \leq \sum_N e^{-\pi A[N]y} \leq \sum_N e^{-\pi cy(n_1^2+\cdots+n_r^2)}$$

$$\leq \left(\sum_N e^{-\pi cyn^2}\right)^r$$

Let $\rho(\nu,A)$ be the number of distinct integral vectors N such that $A[N] = \nu$, that is, the number of ways ν is represented by the quadratic form A. We have clearly

$$\theta_A(z) = \sum_{\nu=0}^{\infty} \rho(\nu, A)e^{\pi i\nu z} \quad .$$

We also have $\theta_A(z+2) = \theta_A(z)$. To show that $\theta_A(z)$ is a modular form, we must study the relation between $\theta_A(-\frac{1}{z})$ and $\theta_A(z)$. To do this, we use the Poisson summation formula.

Poisson Summation Formula: Let $f(x_1,\ldots,x_r)$ be a function of several real variables such that

$$F(x_1,\ldots,x_r) = \sum_{n_1,\ldots,n_r=-\infty}^{\infty} f(n_1 + x_1,\ldots,n_r + x_r)$$

converges absolutely uniformly on compact sets to a differentiable function. Then

$$\sum_{n_1,\ldots,n_r=-\infty}^{\infty} f(x_1+n_1,\ldots,x_r+n_r) =$$

$$\sum_{\nu_1,\ldots,\nu_r=-\infty}^{\infty} e^{2\pi i(\nu_1x_1+\ldots+\nu_rx_r)}\int_{-\infty}^{+\infty}\!\!\ldots\int e^{-2\pi i(\nu_1t_1+\ldots+\nu_rt_r)}\, f(t_1,\ldots,t_r)dt_1\ldots dt_r$$

Proof: This is simply the Fourier expansion of the periodic function $F(x_1,\ldots,x_r)$. (Note that the hypotheses can be weakened considerably).

§20. The Generalized Jacobi Inversion Formula.

Let A be as before and

$$\theta_A(z;X) = \sum_N e^{\pi izA[N+X]} \quad ,$$

so that in particular

$$\theta_A(z;0) = \theta_A(z) \quad .$$

If we apply the Poisson summation formula to $f(t_1,\ldots,t_r) = e^{\pi izA[T]}$ with

$$T = \begin{pmatrix} t_1 \\ \vdots \\ t_r \end{pmatrix}$$

we have

$$F(x_1,\ldots,x_r) = \theta_A(z;x)$$

$$= \sum_M e^{2\pi iM\cdot X}\int_{-\infty}^{+\infty}\!\!\ldots\int e^{\pi i(zA[T] - 2M\cdot T)}\, dt_1 \ldots dt_r \quad .$$

Let us now try to evaluate the integral on the right hand side. In order to do this, we first "complete the square":

$$z(T - \tfrac{1}{z}A^{-1}M) \cdot A(T - \tfrac{1}{z}A^{-1}M) = zT \cdot AT - 2M \cdot T + \tfrac{1}{z}M \cdot A^{-1}M$$

(recall that A is symmetric, so that $X \cdot AY = AX \cdot Y$)

$$= zA[T] - 2M \cdot T + \tfrac{1}{z}A^{-1}[M]$$

(A^{-1} is a positive definite symmetric matrix with rational coefficients).

We ahve therefore:

$$\int_{-\infty}^{+\infty}\cdots\int e^{\pi i(zA[T] - 2M \cdot T)}\, dt_1 \ldots dt_r$$
$$= e^{-\frac{\pi i}{z} A^{-1}[M]}\int\cdots\int e^{\pi i z A[T - \frac{1}{z} A^{-1}M]}\, dt_1 \ldots dt_r \; .$$

Let

$$I(z) = \int\cdots\int e^{\pi i z A[T - \frac{1}{z} A^{-1}M]}\, dt_1 \ldots dt_r \; .$$

<u>Lemma</u> 1. $$I(z) = \frac{1}{\sqrt{(-iz)^r \det A}} \; .$$

<u>Proof</u>: It follows from the definition that

$$I(z) = \int_{-\infty-\frac{1}{z}A^{-1}M}^{\infty-\frac{1}{z}A^{-1}M} e^{\pi i z A[S]}\, ds_1 \cdots ds_r \; .$$

Since the integrand in each plane $u_j = s_j + it_j$ is an analytic function of u_j and

$$e^{\pi i z A[S]} = o(e^{-cs_j^2}) \quad \text{as} \quad |s_j| \longrightarrow \infty$$

uniformly in any finite strip in t_j, we can use Cauchy's theorem to obtain

$$I(z) = \int_{-\infty}^{\infty}\cdots\int e^{\pi i z A[S]}\, ds_1 \cdots ds_r \; .$$

Now $I(z)$ is an analytic function of z on the upper half plane and thus is determined by its value on the imaginary axis,

$$I(iy) = \int_{-\infty}^{\infty}\cdots\int e^{-\pi y A[S]}\, ds_1 \cdots ds_r \; .$$

Since $\pi y\ A$ is a positive definite matrix we can find a real matrix U such that

$$ {}^{t}UAU = I_r \quad \text{(the identity matrix).} $$

Let $S = UT$; then

$$ \begin{aligned} \pi y\ A[S] &= S \cdot A\ S \\ &= U\ T \cdot A\ U\ T \\ &= T \cdot {}^{t}UAU\ T \\ &= T \cdot T = t_1^{\,2} + \cdots + t_r^{\,2} \quad . \end{aligned} $$

Thus

$$ \begin{aligned} I(iy) &= \int_{-\infty}^{\infty} \int e^{(t_1^{\,2} + \cdots\ t_r^{\,2})} \ (\det U)\ dt_1 \cdots dt_r \\ &= (\det U)\ \pi^{r/2} \quad . \end{aligned} $$

From the definition of U, $\det U = \dfrac{1}{\sqrt{(\pi y)^r \det A}}$; so

$$ I(iy) = \frac{1}{\sqrt{y^r \det A}} = \frac{1}{\sqrt{(-i(iy))^r \det A}} \quad , $$

and the proposition follows by analytic continuation.

<u>Generalized Jacobi Inversion Formula</u>:

$$ \theta_A(z;X) = \frac{1}{\sqrt{(-iz)^r \det A}} \left(\sum_M e^{-\frac{\pi i}{z} A^{-1}[M] + 2\pi i X \cdot M} \right) ; $$

and in particular

$$ \theta_A(z) = \frac{1}{\sqrt{(-iz)^r \det A}} \quad \theta_{A^{-1}}(-\tfrac{1}{z}) \quad . $$

<u>Proof</u>: This follows immediately from the lemma and the discussion above.

§21. <u>Representations by Sums of Squares</u>.

Let us consider the special case $A = I$, the identity matrix; then $A[N] = I_r[N] = n_1^{\,2} + \cdots + n_r^{\,2}$. Thus $\rho(\nu, A)$ is the number of ways ν

can be represented as the sum of r squares. Deleting the subscript A, we have seen that this theta series satisfies the equations

$$1.\quad \theta(z+2) = \theta(z)$$

$$2.\quad \theta(-\tfrac{1}{z}) = (-iz)^{r/2}\,\theta(z) \quad ,$$

(the second following from the Jacobi inversion formula). If we take $r = 4k$, then

$$2'\cdot \theta(-\tfrac{1}{z}) = (-1)^k\, z^{2k}\,\theta(z) \quad .$$

In the notation of Chapter V, we have therefore

$$\begin{cases} \theta(z+2)\Big|_k \begin{pmatrix} 1 & 2 \\ 0 & 1 \end{pmatrix} = \theta(z) \quad , \\ \theta(-\tfrac{1}{z})\Big|_k \begin{pmatrix} 0 & -1 \\ 1 & 0 \end{pmatrix} = (-1)^k\,\theta(z) \quad ; \end{cases}$$

so that in general

$$\theta(z)\big|_k L = \chi(L)\,\theta(z)$$

where χ is a multiplicative character for the group G_θ generated by the transformations

$$T^2\colon z \longrightarrow z + 2$$

$$S\ \colon z \longrightarrow -\tfrac{1}{z} \quad .$$

If k is even then the kernel of χ is $G_\theta \supset \Gamma_2$. If k is odd the kernel is a normal subgroup of index 2 in G_θ which does not contain S. Thus in this case the kernel must be Γ_2.

This shows that $\theta(z)$ is an unrestricted modular form of weight k for Γ_2. To show that it is actually a modular form requires a study of the behaviour of $\theta(z)$ at the cusp points of Γ_2 ($i\infty$, 0 and 1).

(1) $\theta(i\infty) = \rho_{4k}(0)$, the number of representations of 0 as a sum of $4k$ squares. Thus $\theta(i\infty) = 1$.

(2) $S(i\infty) = 0$ so that $\theta(0)$ is the value of $\theta(z)|_k S$ (the S-transform of θ) at $i\infty$. Thus $\theta(0) = (-1)^k$.

(3) If $L\colon z \longrightarrow 1 - \frac{1}{z} = 1 + \zeta$, then $L(i\infty) = 1$ and $\theta(1)$ is the value at $i\infty$ of $\theta(z)|_k L$.

$$\theta(z)|_k L = z^{2k}\ \theta(1-\tfrac{1}{z}) = z^{2k} \sum_{m_1,\dots,m_r} e^{\pi i(m_1^2+\dots+m_r^2)(1-\frac{1}{z})} \quad ;$$

since $m_i^2 \equiv m_i \pmod 2$ we can write further

$$\theta(z)|_k L = z^{2k} \sum_{m_1,\dots,m_r} e^{-\pi\frac{i}{z}(m_1^2+\dots+m_r^2) + \pi i(m_1+\dots+m_r)} \quad ,$$

and thus, by Jacobi inversion formula,

$$\theta(z)|_k L = (-1)^k\ \theta(z;X) \quad \text{with} \quad X = \begin{pmatrix} \frac{1}{2} \\ \vdots \\ \frac{1}{2} \end{pmatrix}$$

Since 0 is never represented by $A[N+X]$, N integral, we finally obtain $\theta(1) = 0$.

Recapitulating: the theta function

$$\theta(z) = \sum_{n_1,\dots,n_{4k}} e^{\pi i(n_1^2+\dots+n_{4k}^2)z} = \sum_{n=0}^{\infty} \rho_{4k}(n)\ e^{\pi i n z}$$

is a modular form of weight k for Γ_2. It has the following values at the cusps:

$$\begin{cases} \theta(0) = (-1)^k \\ \theta(1) = 0 \\ \theta(i\infty) = 1 \end{cases}$$

Let $\left\{\begin{matrix} G_k^*(z;0,1;2) \\ G_k^*(z;1,0;2) \end{matrix}\right\}$ be the Special Eisenstein series of weight k, for Γ_2, which is 1 at $\left\{\begin{matrix} i\infty \\ 0 \end{matrix}\right\}$ and vanishes at the other cusps. Then .

$$\theta(z) - G_k^*(z;0,1;2) - (-1)^k G_k^*(z;1,0;2) = \psi(z)$$

is a cusp form for Γ_2 and vanishes identically for $k = 2$ (in that case, there are no cusp forms). A simple calculation using formula (4) of §13 and Theorem 2 of §14 shows that

$$\rho_{4k}(n) = \frac{4k}{(2^{2k}-1)\, B_k} \sum_{\substack{d | n \\ d \equiv n \ (\mathrm{mod}\ 2)}} d^{2k-1} + (-1)^k \sum_{\substack{d | n \\ \frac{d}{n} \equiv 0 \ (\mathrm{mod}\ 2)}} (-1)^d\, d^{2k-1} + a_n,$$

where B_k is the kth Bernoulli number and a_n is the nth Fourier coefficient of $\psi(z)$. As remarked above, we have an exact formula for $k = 2$ ($a_n \equiv 0$); for $k > 2$, we have the weaker result $a_n = O(n^k)$.

§22. Even Integral Quadratic Forms

After our digression of §21 we return to the general theory as in §20. Our final result there was the Jacobi Inversion Formula

$$\theta(z;X) = \frac{1}{\sqrt{(-iz)^r \det A}} \left(\sum_N e^{-\frac{\pi i}{z} A^{-1}[N] + 2\pi i X \cdot N} \right)$$

where

$$\theta(z;X) = \sum_N e^{\pi i z A[N+X]} .$$

We now assume further that

(1) $r = 4k$;

(2) A is <u>even integral</u>, i.e., a_{ij} and $\frac{1}{2} a_{ii}$ are all integers. In particular $A[N] \equiv 0 \pmod 2$ for every integral vector N.

Condition (1) is needed since we have studied only modular forms of integral weight. The other condition is of no function theoretic significance but makes the calculations and results simpler.

A^{-1} is a matrix with rational coefficients, so that we can find an integer q such that qA^{-1} is also even integral; the least such q is called the <u>level</u> of the quadratic form A. (We assume q is odd.)

The main theorem of this section is the following:

<u>Theorem 1</u>: Let X be an integral vector such that $A X \equiv 0 \pmod q$; then $\theta(z, \frac{1}{q} X)$ is a modular form of weight k for Γ_q.

The proof of this theorem is long and so will be decomposed into a sequence of lemmas.

Let $\mathfrak{S}$ be a complete set of distinct representatives $X \bmod q$ such that $AX \equiv 0 \pmod q$. Let N be an arbitrary integral vector; since qA^{-1} is an integral matrix we can write

$$qA^{-1}N = qM + X$$

with M integral and X in $\mathfrak{S}$. Therefore as M and X run through the integer lattice and $\mathfrak{S}$ respectively, $N = A(M + \frac{1}{q}X)$ runs through all integer vectors N in Z^r and the decomposition is unique.

Lemma 2: Let X be in $\mathfrak{S}$ and $S: z \longrightarrow -\frac{1}{z}$; then

$$\theta(z;\tfrac{1}{q}X)\big|_k S = \frac{(-1)^k}{\sqrt{\det A}} \sum_{Y \text{ in } \mathfrak{S}} e^{2\pi i q^{-2} X\cdot AY}\, \theta(z;\tfrac{1}{q}Y) \quad .$$

Proof: The Jacobi inversion formula gives

$$(-1)^k \sqrt{\det A}\;\; z^{2k}\, \theta(z;\tfrac{1}{q}X) = \sum_N e(-\tfrac{1}{z}A^{-1}[N] + \tfrac{2}{q}X\cdot N) \quad ,$$

(where $e(w) = e^{\pi i w}$, so that $e(w + 2n) = e(w)$ if n is an integer)

$$= \sum_M \sum_{Y \text{ in } \mathfrak{S}} e\Big(-\tfrac{1}{z}A^{-1}\Big[\,A(M+\tfrac{1}{q}Y)\,\Big] + \tfrac{2}{q}X\cdot A\,(M+\tfrac{1}{q}Y)\Big) \quad .$$

Unless otherwise specified the summation is over all integer vectors. Since $A^{-1}[AW] = A[W]$, a typical term is

$$e\Big(-\tfrac{1}{z}A\Big[M+\tfrac{1}{q}Y\Big] + \tfrac{2}{q}AX\cdot M + \tfrac{2}{q^2}X\cdot AY\Big) \; ;$$

the middle term is an integer so can be deleted. Interchanging the two summations and using the definitions,

$$(-1)^k\sqrt{\det A}\;\; z^{2k}\,\theta(z;\tfrac{1}{q}X) = \sum_{Y \text{ in } \mathfrak{S}} e(2q^{-2}X\cdot AY)\,\theta(-\tfrac{1}{z};\tfrac{1}{q}Y) \quad .$$

Replacing z by $-\frac{1}{z}$ leads to the desired result.

Lemma 3: Let X be in $\mathfrak{S}$ and $T: z \longrightarrow z+1$; then

$$\theta(z; \frac{1}{q} X)|_k T = e(q^{-2} A[X]) \, \theta(z; \frac{1}{q} X) \quad .$$

Proof:

$$\theta(z+1; \frac{1}{q} X) = \sum_N e((z+1) \, A[N + \frac{1}{q} X])$$

$$= \sum_N e(z \, A[N + \frac{1}{q} X)]) \, e(A[N] + \frac{2}{q} AX \cdot N + \frac{1}{q^2} A[X]) \quad .$$

In the second exponential the first two terms are even integers, since A is even integral and X in $\mathfrak{S}$.

Remark 1: In the proof of Lemma 2 we showed that

$$\theta(z;W)|_k S = \frac{(-1)^k}{\sqrt{\det A}} \sum_M \sum_{Y \text{ in } \mathfrak{S}} e(z \, A[M + \frac{1}{q} Y] + 2W \cdot A(M + \frac{1}{q} Y)) \quad .$$

This result will be needed later on. (W is an arbitrary vector!)

Remark 2: $\theta(z; \frac{1}{q} X)$, with X in $\mathfrak{S}$, is periodic of period q and is holomorphic at $i\infty$, as seen from its Fourier expansion. Its value there is the constant term of the Fourier expansion, which can be interpreted as the number of representations of zero by $A[M + \frac{1}{q} X]$. Thus

$$\lim_{\operatorname{Im} z \longrightarrow \infty} \theta(z; \frac{1}{q} X) = \begin{cases} 1 & \text{if } X = 0 \ , \\ 0 & \text{otherwise} \ . \end{cases}$$

Remark 3: Since the transformations T and S generate the modular group, it follows from Lemmas 2 and 3 that for any L in Γ, $\theta(z; \frac{1}{q} X)|_k L$ is a linear combination of $\theta(z; \frac{1}{q} Y)$ with Y running through $\mathfrak{S}$. This guarantees, in particular, that $\theta(z; \frac{1}{q} X)$ is holomorphic at all cusp points (as a modular form).

In view of the remarks above, we need only show that $\theta(z; \frac{1}{q} X)$ are unrestricted modular forms for Γ_q, to complete the proof of Theorem 1.

Lemma 4:

$$\theta(\zeta + \frac{a}{c} ; \frac{1}{q} X) = \sum_{N \bmod c} e(\frac{a}{c} A[N + \frac{X}{q}]) \, \theta(c^2 \zeta; \frac{N}{c} + \frac{X}{cq}) \quad .$$

Proof:

$$\theta(\zeta + \frac{a}{c} ; \frac{1}{q} X) = \sum_M e((\zeta + \frac{a}{c}) A[M + \frac{1}{q} X]) .$$

Write $M = cL + N$ with L integral and N running through a complete set of residue classes mod c. Then

$$\theta(\zeta + \frac{a}{c} ; \frac{1}{q} X) = \sum_{N \bmod c} \sum_L e(\zeta A[cL + N + \frac{1}{q} X]) e(\frac{a}{c} A[cL + N + \frac{1}{q} X]).$$

The second exponential is independent of L:

$$e(\frac{a}{c} A[cL + N + \frac{1}{q} X]) = e(\frac{a}{c} A[cL] + \frac{2a}{c} cL \cdot A(N + \frac{1}{q} X) + \frac{a}{c} A[N + \frac{1}{q} X]) .$$

The first two terms in the exponential are even integers since A is even integral and X is in $\mathfrak{S}$. (Recall that $A[cL] = c^2 A[L]$). The lemma now follows from the definition of $\theta(z ; W)$, upon rearranging the sums.

Lemma 5:

$$\theta(- \frac{1}{z + \frac{d}{c}} ; W) =$$

$$\frac{(-1)^k (z + \frac{d}{c})^{2k}}{\sqrt{\det A}} \sum_M \sum_{Y \text{ in } \mathfrak{S}} e\Big((z + \frac{d}{c}) A[M + \frac{1}{q} Y] + 2W \cdot A(M + \frac{1}{q} Y)\Big) .$$

This is simply remark 1, with z replaced by $z + \frac{d}{c}$.

Let $R : z \longrightarrow \frac{az + b}{cz + d}$, $c \neq 0$, be a modular transformation. We can write $Rz = \frac{a}{c} - \frac{1}{c^2 (z + \frac{d}{c})} = \frac{a}{c} + \zeta$.

Lemma 6: Let R be as above. Then

$$\theta(z; \frac{1}{q} X)|_k R = \frac{(-1)^k}{c^{2k}\sqrt{\det A}} \sum_{Y \text{ in } \mathfrak{S}} R_{YX} \theta(z; \frac{1}{q} Y) ,$$

with

$$R_{YX} = e\Big(- \frac{2b}{q^2} X \cdot AY - \frac{bd}{q^2} A[Y]\Big) \sum_{N \bmod c} e\Big(\frac{a}{c} A[N + \frac{d}{q} Y + \frac{1}{q} X]\Big) .$$

Proof: From Lemma 4 and 5, together with the comment above, we obtain

$$(-1)^k\, c^{2k}\sqrt{\det A}\ \theta(z;\ \tfrac{1}{q}\,X)\big|_k\, R =$$

$$= \sum_{\substack{N \bmod c \\ Y \text{ in } \mathfrak{S} \\ M}} e\Big(\tfrac{a}{c}A[N+\tfrac{1}{q}X]\Big)\cdot e\Big((z+\tfrac{d}{c})A[M+\tfrac{1}{q}Y] + \tfrac{2}{c}(M+\tfrac{1}{q}Y)\cdot A(N+\tfrac{1}{q}X)\Big)$$

$$= \sum_{\substack{N \bmod c \\ Y \text{ in } \mathfrak{S} \\ M}} e\Big(z\,A[M+\tfrac{1}{q}Y]\Big)\, e\Big(\tfrac{a}{c}A[N+\tfrac{1}{q}X] + \tfrac{d}{c}A[M+\tfrac{1}{q}Y] + \tfrac{2}{c}(M+\tfrac{1}{q}Y)\cdot A(N+\tfrac{1}{q}X)\Big)\ .$$

A straightforward calculation shows that the second exponential is

$$e\Big(-\,\frac{bd}{q^2}\,A[Y] - \frac{2b}{q^2}\,X\cdot AY\Big)\, e\Big(\frac{a}{c}\,A[N + \tfrac{1}{q}\,X + dM + \tfrac{d}{q}\,Y]\Big)\ .$$

Thus

$$(-1)^k\, c^{2k}\sqrt{\det A}\ \theta(z;\ \tfrac{1}{q}X)\big|_k\, R =$$

$$\Big[\sum_{Y \text{ in } \mathfrak{S}} e\Big(-\frac{bd}{q^2}A[Y] - \frac{2b}{a^2}X\cdot AY\Big)\Big]\Big[\sum_{M} e\Big(z\,A[M+\tfrac{1}{q}Y]\Big)\Big]\Big[\sum_{N \bmod c} e\Big(\tfrac{a}{c}A[N+\tfrac{1}{q}X+dM+\tfrac{d}{q}Y]\Big)\Big]$$

In the second sum, for fixed M, $N + dM$ runs through a complete residue system mod c whenever N does, so that that sum is independent of M. Rearranging the sums gives the proposition.

<u>Lemma 7</u>: Let $\begin{pmatrix} a & b \\ c & d \end{pmatrix}$, with $c \equiv 0 \pmod q$, belong to Γ' and

$$G(a,c) = \sum_{N \bmod c} \epsilon^{\frac{\pi i a}{c} A[N]}\ .$$

Then if X, Y belong to $\mathfrak{S}$

$$H(X,Y) = \sum_{N \bmod c} e\Big(\frac{a}{c}\,A[N + \tfrac{d}{q}\,Y + \tfrac{1}{q}\,X]\Big) = \begin{cases} G(a,c) & \text{if } X = -dY\ , \\ 0 & \text{otherwise}\ . \end{cases}$$

<u>Proof:</u> Since q is the level of A and $c \equiv 0 \pmod q$, cA^{-1} is an even integral quadratic form. Thus, for any L in Z^r, $cA^{-1}L$ is in Z^r; and as N runs through a complete residue system mod c, so does $N + cA^{-1}L$. Hence

$$H(X,Y) = \sum_{N \bmod c} e\Big(\frac{a}{c}\,A[cA^{-1}L + N + \tfrac{d}{q}\,Y + \tfrac{1}{q}\,X]\Big)\ ;$$

$$= \sum_{N \bmod c} e\left(\frac{a}{c} A[N + \frac{d}{q} Y + \frac{1}{q} X]\right) e\left(2 \frac{a}{q} (X + dY) \cdot L\right)$$

$$= e\left(2 \frac{a}{q} (X + dY) \cdot L\right) H(X,Y) .$$

Thus $H(X,Y) = 0$ unless $a(X+dY) \cdot L \equiv 0 \pmod q$ for all L in Z^r or $a(X+dY) \equiv 0 \pmod q$. Since $ad - bc \equiv 1$ and $c \equiv 0 \pmod q$, a and q are coprime and thus $X + dY \equiv 0 \pmod q$. This will occur precisely once since X and Y are coset representatives mod q. In that case, $X + dY = qK$ with K integral, so

$$H(X, -dX) = \sum_{N \bmod c} e\left(\frac{a}{c} A[N+K]\right) = G(a,c) .$$

<u>Corollary</u>: Let $R = \begin{pmatrix} a & b \\ c & d \end{pmatrix}$ belong to Γ_q, and $c \neq 0$; then:

$$\theta(z; \frac{1}{q} X)\Big|_k R = \frac{(-1)^k G(a,c)}{c^{2k}\sqrt{\det A}} \theta(z; \frac{1}{q} X) .$$

<u>Proof</u>: This is a consequence of Lemma 6 and 7, upon noticing that

(1) $\theta(z,W) = \theta(z, -W)$;

(2) $\theta(z, \frac{1}{q} W) = \theta(z, \frac{1}{q} Z)$ if $W \equiv Z \pmod q$.

<u>Remark 4</u>: If $c = 0$, then $R = \begin{pmatrix} 1 & bq \\ 0 & 1 \end{pmatrix}$ and

$$\theta(z; \frac{1}{q} X)\Big|_k R = \theta(z, \frac{1}{q} X) ,$$

as follows from Lemma 3.

<u>Lemma 8</u>: Let $(a,c) = 1$, $R = \begin{pmatrix} a & b \\ c & d \end{pmatrix}$; then

$$\theta\left(\frac{a}{c} ; \frac{1}{q} X\right) = \frac{\varphi(a;c|X)(-1)^k}{c^{2k}\sqrt{\det A}} ,$$

where

$$\varphi(a;c|X) = \sum_{N \bmod c} e\left(\frac{a}{c} A[N + \frac{1}{q} X]\right) .$$

<u>Proof</u>: By definition

$$\theta\left(\frac{a}{c} ; \frac{1}{q} X\right) = \lim_{z \to i\infty} \theta(z; \frac{1}{q} X)\Big|_k R ;$$

and by Lemma 6, then

$$(-1)^k c^{2k}\sqrt{\det A}\quad \theta\left(\frac{a}{c}\,;\,\frac{1}{q}X\right) =$$

$$= \sum_{\substack{N \bmod c \\ Y \text{ in } \mathfrak{S}}} e\left(\frac{a}{c}A[N+\frac{1}{q}X+\frac{d}{q}Y] - \frac{b}{q}(2X\cdot AY + dA[Y])\right) \lim_{z\to i\infty} \theta(z;\frac{1}{q}Y) \quad .$$

The proposition now follows by remark 2.

Lemma 9: (Generalized reciprocity theorem) Let $ad - bc = 1$ as before; then, in the notation of Lemma 8,

$$\frac{(-1)^k\sqrt{\det A}}{c^{2k}}\quad \varphi(a;c\,|X) = \frac{1}{d^{2k}}\sum_{Y \text{ in } \mathfrak{S}} \varphi(b;d\,|X+cY)\; e\left(\frac{a}{q^2}(2X\cdot AY - cA[Y])\right) \quad .$$

Proof: Let $R = \begin{pmatrix} a & b \\ c & d\end{pmatrix}$; then

$$\theta\left(\frac{a}{c}\,;\,\frac{1}{q}X\right) = \lim_{z\to i\infty} \theta(z;\frac{1}{q}X)\Big|_k R$$

$$= \lim_{z\to i\infty} \theta(z\,;\frac{1}{q}X)\Big|_k RS\Big|_k S \quad ,$$

where $S: z \longrightarrow -\frac{1}{z}$, $RS = \begin{pmatrix} b & -a \\ d & -c\end{pmatrix}$. Applying Lemma 6 to $\theta(z\,;\frac{1}{q}X)\Big|_k RS$ gives

$$(-1)^k d^{2k}\sqrt{\det A}\quad \theta\left(\frac{a}{c}\,;\,\frac{1}{q}X\right) =$$

$$= \sum_{Y \text{ in } \mathfrak{S}} e\left(-\frac{a}{q^2}(2X\cdot AY - cA[Y])\right)\varphi(a;c\,|X+cY) \lim_{z\to i\infty} \theta(z\,;\frac{1}{q}Y)\Big|_k S \quad .$$

The proposition now follows from Lemmas 2 and 8, and Remark 2.

Corollary (Reciprocity Law for Gaussian Sums). Let $c \equiv 0$ (mod q); then if q is odd

$$\frac{(-1)^k\, G(a,c)}{c^{2k}\sqrt{\det\ A}} = \frac{G(b,d)}{d^{2k}} \quad .$$

Proof: We note that $\varphi(a;c\,|0) = G(a,c)$, and substitute $X = 0$, $c \equiv 0 \pmod q$ in Lemma 9 to obtain

$$\frac{(-1)^k\sqrt{\det A}}{c^{2k}}\, G(a,c) = \frac{G(b,d)}{d^{2k}}\sum_{Y \text{ in } \mathfrak{S}} 1 \quad .$$

(Recall that $\varphi(a,c\,|X) = \varphi(a,c\,|Y)$ if $X \equiv Y \pmod c$). It follows from the discussion before Lemma 3 that the set $\{\frac{1}{q}(X)\,|X \text{ in } \mathfrak{S}\}$ is a complete set of coset representatives modulo $A(Z^r)$. Since the index of $A(Z^r)$ in Z^r is $\det A$, the corollary is proved.

"Corollary to the Corollary:" Under the same conditions as above,

$$\frac{1}{d^{2k}} G(b;d) = \frac{1}{(d+\ell c)^{2k}} G(b+\ell a; d+\ell c)$$

for any integer ℓ.

Proof: The corollary can be applied both to $\begin{pmatrix} a & b \\ c & d \end{pmatrix}$ and to $\begin{pmatrix} a & b+\ell a \\ c & d+\ell c \end{pmatrix}$. Comparison leads to the desired result.

Lemma 10: The Gaussian sum $G(b;d)$ is a rational number. Moreover, $G(b;d) = G(1;d)$.

Proof: By definition,

$$G(b;d) = \sum_{N \bmod d} e^{\pi i \frac{b}{d} A[N]} .$$

Since A is even integral, $G(b;d)$ is in the field generated by the dth roots of unity. The corollary to the corollary shows that it is also in the field by the $(d+\ell c)^{th}$ roots of unity for all integers ℓ. Since c and d are coprime, so are d and $d + \ell c$, for some ℓ. Thus $G(b;d)$ is rational and therefore invariant under all automorphisms of the field of dth roots of unity. In particular, using the automorphisms generated by sending $e^{\pi i b/d}$ into $e^{\pi i/d}$, it follows that $G(b;d) = G(1;d)$.

Remark: In the proof above, we used a result from the elementary theory of fields. Let F_m be the field of mth roots of unity; then if $(m,n) = 1$, $F_m \cap F_m = Q$, the field of rational numbers. In this case F_{nm} is the smallest field generated by F_n and F_m; to obtain the result, one compares the degrees of the extensions involved.

Corollary: Let $d \equiv 1 \pmod q$; then

$$H(d) = \frac{G(d;d)}{d^{2k}} = \frac{G(1;d)}{d^{2k}} = 1 .$$

Proof: The "corollary to the corollary" give

$$H(d) = H(d+\ell c)$$

under the assumption that $c \equiv 0 \pmod q$. Let $c = q$, a and b be chosen so that $ad - bq = 1$. Thus $H(d)$ depends only on the congruence class of $d \bmod q$. Since $d \equiv 1 \pmod q$ and $H(1) = 1$, we are through.

Finally, we have proved Theorem 1. This is a consequence of the corollaries to Lemmas 7, 9, and 10.

§23. Arithmetic Applications.

Let $\theta(z) = \theta(z;0) = \sum_N e^{\pi i A[N]z}$, where A is, as usual, an even integral, positive definite, symmetric quadratic form of odd level q. We have seen, in Theorem 1, that $\theta(z)$ is a modular form of weight k (the dimension of A being $4k$) for the principal congruence subgroup Γ_q. We have also evaluated $\theta(z)$ at the cusp points. Thus we can subtract a suitable linear combination of Eisenstein series to obtain a cusp form. Comparing coefficients leads to good asymptotic results on $\rho(n;A)$, the number of representations of n by A.

We shall restrict our attention now to the case $q = 1$.

Theorem 2: If the quadratic form A in r variables is of level 1, then $r \equiv 0 \pmod 8$.

Proof: In this case the Jacobi inversion formula reads as follows:

$$\theta(z)\Big|_{\frac{r}{4}} S = i^{\frac{r}{2}} \sum_N e^{\pi i z A^{-1}[N]}, \qquad (S: z \longrightarrow -\tfrac{1}{z}) .$$

Since $q = 1$, $A(N)$ runs through all integer vectors as N does. Thus, since

$$A^{-1}[A\,N] = A[N] ,$$

$$\theta(z)\Big|_{\frac{r}{4}} S = i^{r/2}\,\theta(z)$$

Since $S^2 = I$, $i^r = 1$, so that $r = 4k$. But in this case Theorem 1 tells us that $\theta(z)$ is a modular form of weight k for Γ and so $\theta(z)\big|_k S = \theta(z)$. We have $i^{r/2} = 1$ and therefore $r \equiv 0 \pmod 8$.

If A is a quadratic form of level 1 in $4k$ variables (k even), then $\theta(z)$ is a modular form of weight k for Γ and has the value 1 at $i\infty$. Since the normalized Eisenstein series

$$E_k(z) = 1 + \frac{(-1)^k\, 4k}{B_k} \sum_{n=1}^{\infty} \sigma_{2k-1}(n)\, e^{2\pi i n z}$$

has the same properties,

$$\theta(z) = E_k(z) + c(z)$$

where $c(z)$ is a cusp form of weight k. Comparing coefficients gives

$$\rho(2n;A) = \frac{4k}{B_k}\ \sigma_{2k-1}(n) + O(n^k) \quad ,$$

and for $k = 2,4$ (in which case there are no cusp forms) we have the exact formulas

$$\rho(2n;A_8) = 240\ \sigma_3(n) \quad ,$$

$$\rho(2n;A_{16}) = 480\ \sigma_7(n) \quad .$$

For examples of such quadratic forms, the reader is referred to Hecke's works [2].

BIBLIOGRAPHY

[1]. Eichler, M , Quadratische Formen und Orthogonale Gruppen, Springer, (1952).

[2]. Hecke, E., "Analytische Arithmetik der positiven quadratischen Formen," Kgl. Danske Videnskabernes Selskab. Mathematisk- fysiske Meddelelser, XVII, 12, (1940).

[3]. Schoeneberg, B.S., Math. Ann., vol. 116 (1938), pp. 511-523.

[4]. Siegel, C. L., Annals of Math., vol. 36 (1935), pp. 527-606; vol. 37 (1936), pp. 230-263; vol. 38 (1937), pp. 212-291.